THE DISCIPLINE OF CURIOSITY

For
Otto ter Haar
Publisher

THE DISCIPLINE OF CURIOSITY

Science in the world

Janny Groen, Eefke Smit, Juurd Eijsvoogel
Editors

ELSEVIER SCIENCE PUBLISHERS
AMSTERDAM, LONDON, PARIS, NEW YORK, TOKYO
1990

Published by:

ELSEVIER SCIENCE PUBLISHERS B.V.
Sara Burgerhartstraat 25
P.O. Box 211
1000 AE Amsterdam
The Netherlands

Distributors for the United States and Canada:

ELSEVIER SCIENCE PUBLISHING COMPANY, INC.
655 Avenue of the Americas
New York, N.Y. 10010
U.S.A.

Cover design: Nigel Hynes

Library of Congress Cataloging-in-Publication Data

The Discipline of curiosity : science in the world / Janny Groen,
  Eefke Smit, Juurd Eijsvoogel, editors.
      p.   cm.
    Includes index.
    ISBN 0-444-88861-6
    1. Science--Social aspects. 2. Research--Social aspects.
  I. Groen, Janny. II. Smit, Eefke. III. Eijsvoogel, Juurd.
  Q175.55.D57   1990
  306.4'5--dc20   90-13974
                                                          CIP

ISBN: 0 444 88861 6

PRINTED IN THE NETHERLANDS

Contents

Introduction

In the 20th century, more than ever before, the world is being shaped by science. The scientific enterprise is not only "millions of people trying to understand how the world ticks," to quote John Maddox, the Editor of 'Nature'. It is much more. Science is providing today's society with the kind of information that will change the society of the future. Scientists have become leading actors in world history.

The results of scientific research determine the way food is produced, wars are fought and procreation is regulated. Communication, health care, transport, trade, industry and countless other aspects of daily life have been drastically and permanently changed as a result of the discoveries made by that disciplined form of human curiosity that we call science.

Science is, in the words of the American author David Halberstam, 'the engine of modern society'. "It is evident that we live in a technology-driven world and that those societies who excel at science, not just pure science but the application of it, are going to be the successful ones." Knowledge, primarily scientific knowledge, provides the new raw material for prosperity. It is a major instrument, to some even a weapon, in the rivalry between social classes, countries and trading blocks.

For some time now, scientists have no longer been able to shut themselves away in their proverbial ivory towers. Even in the socialized universities, where they could still shelter from the harsh world, in the last decades it has become clear that there is no way in which they can shirk their prominent role in the society of the late twentieth and early twenty-first century.

Science has a central place in today's world. And if a rare scientist does not yet realize this, politicians, activists, journalists, economists and lawyers are ready to remind him of it. Science is in the world, that is to say, science can be found at school, on Wall Street, in the home, in the factory, on the battlefield and in the news.

In this book of interviews, fifteen opinion leaders (prominent figures from international politics, business, communication and science) give their vision of the changing role of science in society. For with the recognition of its crucial importance, comes the understanding that the scientific enterprise is, or should be, a concern to everybody.

Science has an intrinsic value, as the mathematician Roger Penrose stresses in this book, "comparable to the value of artistic things". This may be true, but science obviously has an enormous and increasingly important social value too. The discipline of curiosity, as science may be called, is not just a discipline of form, of methods and means, it is also a discipline of ends; political, moral, cultural and commercial ends.

In all its diverse forms, science, therefore, is increasingly the subject of discussion in parliaments, board meetings, protest demonstrations, diplomatic circles, newspaper articles ... and in the academic world itself. Science and technology have become features of daily life. But we ask ourselves: how does the technological expansion fit into our way of life? Science and its opportunities, its dangers and objectives have different values for different groups in society. The development of genetic mapping and the application of gene-splicing in biotechnology, for example, raise ethical questions. Science has become a matter of political debate and of the law. But science is also an industrial and commercial activity, and the scientific objectives of governments and industry may differ as much as their respective responsibilities.

All the people we asked to reflect on the role of science in the world, or in their world, agreed on the importance of the matter:
– David Halberstam, the American journalist and author of 'The Best and the Brightest', 'The Powers that Be' and 'The Reckoning';
– Seun Ogunseitan, the Nigerian journalist who made the world press and who is now setting up a system for the distribution of scientific information in a part of the world that tends to be forgotten by some above the equator;
– Federico Mayor Zaragoza, a former Professor of Biochemistry who is now Director General of Unesco;
– Alexander King, one of the founders of the Club of Rome and currently its President;
– Erich Bloch, the controversial Director of America's National Science Foundation;

– Harry Beckers, the top scientist of Royal Dutch/Shell, one of the world's largest companies;
– Etienne Davignon, a former Vice President of the European Commission who is currently Chairman of the Belgian conglomerate Société Générale de Belgique;
– Robert Solow, the American economist who won a Nobel Prize in 1987;
– Hisao Yamada, Director of Research and Development of the National Center for Science Information System (NACSIS) in Japan, and Professor of Information Science and Management at the University of Tokyo;
– Tudor Oltean, a former Romanian refugee, now a Professor of Communication Sciences at the University of Amsterdam;
– Rudolf Bernhardt, one of Europe's most experienced and influential judges, who is Director of the Max Planck Institute of Comparative Public Law and International Law in Heidelberg, and representative of the Federal Republic of Germany in the European Court of Human Rights in Strasbourg;
– Roger Penrose, the Oxford Professor of Mathematics who discovered the Penrose Tiles and wrote 'The Emperor's New Mind', strongly opposing the idea that computers can replace human thinking;
– Kai Siegbahn, the Swedish physicist awarded a Nobel Prize in 1981;
– Shigeo Minowa, the Japanese economist who is Director of the Institute of International Business and Management at the Kanagawa University in Tokyo;
– John Maddox, the Editor of the influential science journal 'Nature', and the author of a number of science-related books, like 'The Doomsday Syndrome' and 'Beyond the Energy Crisis'.
They were interviewed by Juurd Eijsvoogel, Janny Groen, Eefke Smit, Anita de Waard, Mari Pijnenborg and Bernard Dixon.

Science will play an increasingly important role in politics all over the world, says Federico Mayor. "Every day we see a further scientification of the decision making process. And I am convinced this will continue."

If the decision makers want to base their policy on scientific data, as he implies, they have to have access to the enormous reservoir of scientific knowledge in the world. But they do not speak the language of science, anymore than the scientists can express themselves in the language of the decision maker. Thus, the communication between science and society remains defective.

Scientific information is essential, not only for the scientist. The politician, the entrepreneur and the public at large need to know about it too. The people in business find that neither the mass media nor the specialized scientific press are providing the information needed. General information is no longer enough, specialist information is only digestible for the learned. Who will bridge the gap?

A society in which all sorts of information are available in greater quantities than ever before, appears to have an urgent need for information that is adequate, that is comprehensible to the various groups of potential users. The value of information lies in its capacity to be communicated and this can be maximized by expanding the number of people that are willing and able to grasp what it means. As long as the supply of adequate scientific information remains stagnant in most parts of society, we are wasting time, money and creativity. Apart from the scientists themselves, publishers, journalists, consultants, documentalists and others in the communication and information industry must ensure that knowledge and information are not only readily available, but also in an understandable, accessible and useable form. Only then can information reach its full potential, which lies in communication.

The output of science is "a highly important kind of commodity," says Erich Bloch. "It has to find its way into the economy of a country." According to Bloch our society is going through a complete restructuring, "in as much as each country and every economy will be much more dependent on knowledge than previously. They used to be highly dependent on natural resources. Now, the economic competitiveness of a country will depend more on human resources, on people." Economic power, in other words, will come increasingly from the maximization of the human intellect.

In such a world, science, information and education are of the utmost importance, for the individual and for the local, national and supra-national communities. Parents do everything possible to enrol their children in a good kindergarten, fearing that otherwise they will not be admitted to the best primary school, and thus not to the best secondary school and thus not to the best university.

Similarly, industries, countries and trading blocks increasingly realize that good education and research are essential for economic progress.

The economic theory of comparative advantage has been revised in line with these views by Robert Solow, who gives a prominent role to education, training and the creation of high technology. "When I was a student, the assumption was that a nation's comparative advantage depended mainly on its natural resources, climate, location, nearness to raw materials and markets." Today comparative advantage is something that a nation or an industry creates for itself. This can be achieved by learning, learning from other people, other companies or other countries.

But not everyone is able to participate in this often very costly race for comparative advantage. The danger of an unequal spread of knowledge is hardly a new phenomenon, but the very fast developments in science and technology make the situation for those who lag behind more and more hopeless. In spite of many efforts to bridge the knowledge gap on a national as well as a global scale, it is widening ever faster.

Some fear that society may be polarized along the lines of education, that there will be people who are technically skilled and those who are not. These technically illiterate men and women will be unable to form their own opinions, they will no longer play a part in the democratic process, because they have not learnt to understand. They may simply look at science as the only way to solve all problems. And that, of course, creates a problem because science is not almighty.

Developing countries often lack the most basic instruments for science and education. School children have to work without textbooks, scientists without up-to-date publications and libraries without an acquisition budget. African scientists laugh bitterly at the idea of being called scientists. Neither they nor their universities can afford to subscribe to scientific journals or professional bulletins published outside the country. "Many people in our universities," Seun Ogunseitan says, "are not sure what is the state of science." They have to rely on what they are told by the national press, friends or 'Time' magazine.

Many people interviewed in this book are highly concerned about the imminent danger posed by the ever widening knowledge

and technology gap between the industrialized countries and the
Third World. Everyone and every country should be able to tap
into science sources, Solow believes. For poor countries too it is
essential to have scientists, according to Mayor. Halberstam predicts serious political tensions between the industrialized countries of
the North and the developing countries in the South because of the
technology gap. "The North has power over the information channels, has access to education and controls the size of its population."
Researchers in developing countries who do not have access to
information about new developments in science, feel isolated from
the rest of the world and cut off from the future. The highly
internationalized science of today is still far from a worldwide
enterprise.

Intensive scientific communication does exist between Europe
and the United States on the one hand and Europe and Japan on
the other. But a peculiar division of labour has established itself
between East and West, a silent arrangement in which the United
States and Europe do the basic science, and Japan turns their
discoveries into marketable products....

Different cultures have different approaches to science. Really
new and unorthodox ideas, so important in the creative part of the
scientific enterprise, are not always well received in Japanese society.
The education system is not stimulating creativity or curiosity in
children. The requirement of uniformity is much more important in
Japan.

But science is not just curiosity and creativity; it is a disciplined
form of it. And that self-same discipline may be what is lacking in
the West; the discipline to follow up on the brilliant new ideas with
down-to-earth applications, the discipline to spend a lot on R & D,
to work hard and to recognize the primacy of education. It is the
seemingly odd couple of discipline and curiosity that makes for
scientific progress. The curiosity of creative minds, asking continuously How? and Why?, and the discipline to realize that science is
part of the world, that it is shaping it, for better and for worse.

The scientist is standing in the middle of the world and has to
communicate with it. The politician, the business person, the activist,
the journalist/observer and the scientist all agree on that. But how?
Many views are given on that unsolved problem in this book, and
much more can and will be written about it. But while the role of

science in the world is changing, and with it the responsibilities of the scientist, one thing does not change. To quote Siegbahn: "In order to get new ideas, you have to read."

Juurd Eijsvoogel

DAVID HALBERSTAM, the observer

The haves and have nots

Science as such does not interest writer, journalist and observer David Halberstam, but science as a phenomenon of modern society does. One of his distinctive talents as a writer is his ability to connect the course of a man's career with the culture, society and institutions that surround and shape him. The themes that attract him – ambition, career, power – are the themes of his life and his books.

Halberstam is a native New Yorker and a 1955 graduate of Harvard College. Fresh from the university, he went to work as a reporter on the smallest daily newspaper in Mississippi. He then spent four years on the Nashville 'Tennessean', and for six years served as a foreign correspondent for the 'New York Times' in the Congo, Vietnam and Poland. He had written several books before 'The Best and the Brightest', his study about the US Government and military in the Vietnam era that established his reputation. Other major works followed: 'The Powers that Be', about the rise of modern media; and 'The Reckoning', a dual history of the Ford and Nissan automobile companies. His latest work, "a little booklet, something between a book and a magazine," he says, is about the year 2000 and beyond called 'The Next Century'.

"We live in a technology driven world and those who excel at science, not just pure science, but also its application, are going to be the successful societies," states Halberstam. "Japan is a self-evident example. On the other side is the dilemma now faced by the Soviet Union, a controlled society fearing all the inventions of modern technology: the Xerox machine and the computer. Can it be afraid of these machines and still be competitive? The Soviet Union has to open up and that's what the Gorbachev revolution is all about. Competitiveness and repression do not go together. The consequence of repression is to stunt essential growth, not just to diminish the Gross National Product, but also to weaken national security. So, those who are good in modern technology are going to be the winners, and those who are not are going to fall behind."

Maximization of the human intellect: the new economic power

"Is there a new definition of economic power?," he queries. "The American century of economic power derived from great natural resources, a good climate, inexpensive energy, and large-scale production has come to an end in the last ten years or so. I won't call the next century the Japanese or the Asian century, although they are the pioneers. In this century economic power will come from the maximization of the human intellect."

"In a sense Japan is an odd country for an economic power. It is a rather small nation, living on small, inhospitable, rocky islands with very little agrarian land, no natural resources, and nothing under the soil. So they've achieved economic power by educating their people, by pulling them into science education and flooding the factory floor with engineers. This is the new definition of economic power; and nations wanting to compete with them have to meet this challenge."

"Many people still get it wrong," he sighs. "They want to believe that the Japanese have succeeded because they had a soft yen or a cheap labour force. Education and high science are the driving components of that society. Japan is going forward now and is re-investing in research and development at a higher rate than we are in the USA."

"It is a great mistake to think of the Japanese as mere imitators. Certainly, in the last twenty or thirty years they have been good at taking other people's inventions and applying them better. In many areas, however, they are now becoming the initiators. They are producing scientists at twice the per capita rate than we are. Over a period of say five, ten or twenty years, there are benefits to be reaped. Nobody can deny that Japan is a greatly talented society."

Halberstam does not believe that the Japanese, with their maximization of the human intellect, will be the new rulers of the world. "They are an economic power as yet unmatched by political military and diplomatic power," he asserts. "It is an industrial power, which is, for the first time, being matched by financial power. Because they work so hard they are liquid and thus able to invest in research and development, and at the same time are beginning to buy the things which will bring some power. There are great limits to Japan becoming a hegemonic power: their natural, cultural protectiveness irritates some, and their language is a barrier."

"There is a contradiction between their economic and military power. Japanese economic power comes in no small degree because it has not had a military budget as have the United States and the Soviet Union. A military budget would have immediately reduced their economic power, and they would not have had a monetary surplus to invest in education, research and development. Japan is likely to be a powerful force in the next forty or fifty years."

Primacy of education: the driving force

The driving force is the primacy of education, is Halberstam's opinion. "I regard Japan as the culture of adversity whereas the United States is the culture of affluence. With so little in the way of natural resources they have had to maximize their system, and in doing so a society which does not waste has emerged. And they have had this bountiful, endless fruitful society for three or four generations, while we in the United States have built into our culture expectations that can no longer be delivered."

He agrees that American universities still produce world renowned scientists; that they are among the best in the world. "We may have the greatest higher education plants, but only a small percentage of native born Americans are involved in their operation. The number of foreigners, particularly Asians, is increasing in our graduate schools, especially in the scientific plants, and they are educating the rest of the world. But this could change. We could have a government that recognized the signs of our malaise and set out incentives to encourage our young people to be scientists instead of going to business schools, in which case the situation could turn around very quickly."

Should the United States change its priorities, and invest more in scientific research and development? "Of course, for the sake of our children, our nation and for the world," is the answer. "A strong America is a sign of hope for the rest of the world. Without meaning to be chauvinistic, it seems to me that with the exception of the Vietnam war, by and large American power has been a relatively stabilizing factor in the world in the post-Second World War period. A strong and economically healthy America is a positive influence in global terms."

"We are still pretending that we are living in a position of American economic hegemony, and have not adapted our political system to the new circumstances of an international economy, in which Japan and West Germany are good and getting better. The events in Eastern Europe make me think that Germany will become the next Japan. Yet in the United States, we are going on as if it was still 1955 and as if we are rich in a poor world."

The future of the United States is not just a matter of investing in research and development. America is unlikely to fall far behind in science and technology, considers Halberstam. "While we may need the help of foreigners, at the high education level we'll continue to be good. The question is whether we will continue to lose our blue collar industrial and middle-class jobs. In the past a good blue collar worker did not need much education. Now in this computer driven society, he needs a high school education and maybe two or three years in the Air Force. We are not competitive here. At the high level, because of our great scientific component, we are still competitive and whether at the middle level is the question."

Science and technology: North-South tension

In a world less plagued by East-West military tensions, will a new type of war emerge, science and technology wars? Halberstam does not think so. "Nobody really knows what is going to happen; situations change so quickly. The Soviet Union seemed to be heading for an endless repressive rule in Eastern Europe and then suddenly gave up its empire, without demanding anything of the West. Now that Europe is going to redefine itself, world tensions will no longer be East-West, but increasingly North-South, between the 'haves and have nots'.

"The North has access to science and technology and is wealthier. I am afraid that the increasing importance of science and technology will lead to new tensions. The North has power over the information channels, has access to education and controls the size of its population. I think that the rich will get richer, and that's a serious problem."

"In considering the tensions between the North and the South, the rich and the poor, a distinction needs to be made between those who control their population and those who do not. Why has Japan been successful? One of the keys to their economic success is that they have dealt with the problem of overpopulation. Japan was the first non-white country after World War II to introduce birth control effectively. Japan must never be thought of as an underdeveloped country; it is an Asian nation. The message is clear for the underdeveloped countries: maximization of the human intellect is virtually impossible in an overpopulated nation. The first step thus is to apply birth control."

In the tension between North and South, the capacity of the people in the South to have a primitive atomic bomb is a real danger. In expressing his concern, Halberstam goes on to say, "We can't do anything to prevent them from making such a bomb. The number of nations having atomic weapons is increasing all the time, and this is a real threat for the next ten, fifteen years. An atomic bomb is not hard to make in the contemporary world. A young Soviet journalist was saying that one of the dangers for the Soviet Union is that the people in Azerbaijan or Baku could make an atomic bomb."

Science and technology: class-based society

"Science and technology in the new post-industrial state is creating a more class-based society, especially in the United States and Japan. In the old industrial state, a blue collar worker in the modern era could get virtually a middle-class wage and end up with a house or apartment, a car, eat well, put protein on the table, and send his children to college. In a society where the demarcation

point is increasingly science and technology, the blue collar part of
the economy will continue to decline. These jobs will be exported
from the United States and also Europe to poorer countries in Asia
and Latin America, or will become super-automated."

"Will the middle class diminish to a small group of winners who
do very well, and an increasing number of people working in the
low service economy and in jobs without any dynamism? This is
happening in Japan and in the United States now. The intensity is
to get a child into a good school early, because this will lead to a
good college, which in turn will lead to a good job. For my
daughter's entry into private school, a friend, a book editor, wrote a
letter of application which began, 'While I have never written a
letter of recommendation for a four-year old...'."

Halberstam does not believe that a new powerful elite of scientists will emerge to rule the world. "Economies will be science
driven and those societies which understand that science is the
engine of a modern society can have certain modern military
strength. As Gorbachev realizes, military strength is not enough. He
understands that the Soviet Union is falling behind not just United
States and Japan, but also South Korea. As I said at the beginning,
if you are afraid of the Xerox machine, how can you get into the
world of the computer?"

"They have a computer system in schools in the Soviet Union,
but it doesn't work. The children have to write down what they did
on a piece of paper, because their computers can't store the information. Every child in Japan and the USA has access to a computer. An interesting aspect of modern inventions is that they are,
among other things, inventions of communication. Being afraid of
modern communication, is like shooting off your toes. My daughter,
who is nine years old, and most of her classmates, have easier access
to a modern computer than most people in the Soviet Union."

Should we share high tech information with countries in Eastern
Europe? "Yes," says Halberstam, "sharing information is healthy.
You don't give away levels of high technology where you make
breakthroughs, not if you want to keep the competitive edge. We
ought to improve and create more information channels between
East and West, between North and South, between governments,
private organizations and scientists. If we had shared information
with the Soviets, they might not have gone into Afghanistan."

"Obviously we should share information of benefit to mankind. Pollution and diseases like AIDS are world problems and cooperation should be encouraged between scientists working on problems such as an AIDS pill or the greenhouse effect. If someone finds a solution to a world problem, a certain degree of commercialism may be involved, but the world will not tolerate exclusivity. As with the birth control pill, an AIDS pill would need to be distributed as widely as possible in countries in Eastern Europe, Asia, Africa and Latin America. The rationale of humanity will overwhelm any selfishness. The people in the pharmaceutical company would give away the secrets to some other company or leak it to the press. In that sense, I think the media have an important role to play."

Science in the media world

"Yes, science and technology have changed the nature of power. We are living in a media world. Former US Secretary of State George Schultz once said: 'In the past a nation was rich if it had a lot of copper under the soil. Today we live in an age of fibre optics. We have satellites; we can watch the Germans tearing down the Berlin Wall. Technically the world has shrunk'."

"Thirty years ago I would have said television is going to make the world smaller and more international. But because of commercial norms, American television is virtually isolationist. The evening news shows very little of foreign events, mostly a bombing or killing but no evolutionary news from a foreign country. What should be a force for internationalism has in reality become an isolationist force. The journalists may carry very useful information, but it's only worth something if people are ready to listen. Most people are not ready to listen, especially when it concerns very complicated information. This has to do with the poor standard of education. So we are back to the key to a successful nation: education, maximization of the intellect."

While information is a vital part of modern society, Halberstam is not afraid of terrorists or hostile countries invading information channels, stealing secrets, disrupting information traffic, or paralysing telephone networks. "One day a terrorist organization will probably take over a major television network, or spread viruses in computer networks. Such acts may be exciting for a day or two but life goes on."

"Do I know who is controlling who in this high tech world? I don't even know what the word control means," he muses. "The television networks are very powerful, but it is also true that programme producers are responding to what they think viewers want. They assume that audiences do not want to see things about Europe or science. So are they controlling the people or are the people controlling them?"

"An interesting aspect of recent events in Eastern Europe is the defeat of the Orwellian image from 1984, of the totalitarian state controlling its people by technology, that is, communication, television. In all these countries for thirty or forty years, the only control of radio, press and television has been the state, and it has not worked."

"Everybody in these Eastern European countries has seen a movie or television programme from somewhere else, read a book, or has a cousin in America or Europe. People's desire to be free, to be curious, can not be suppressed. Ideas can not be controlled. Voltaire once said: 'There is nothing more powerful than an idea whose time has come,' and I think that's true in today's era of high science."

Janny Groen

The growing tension between North and South, between the haves and the have nots and between those who have access to information and those who do not, is one of Halberstam's main concerns. The Nigerian journalist Seun Ogunseitan has similar concerns. He has had ample opportunity to become familiar with the magnitude of the problem. Firstly as a student of zoology at the University of Ibadan, and then as the Science Editor of the respected Nigerian newspaper 'The Guardian'. In Nigeria, where one-fifth of the total African population lives, the shortage of information is acute, Ogunseitan says. School children have to do without textbooks, scientists without up-to-date publications and university libraries without an acquisition budget.

SEUN OGUNSEITAN, the information gatherer

An African dream

Oluwaseun Oladapo Ogunseitan joined the staff of 'The Guardian' as a science correspondent in 1984. Born in 1960, this young man specialized in reporting on the environment, in those days a rather new field of interest for the press. A news story about the pollution caused by a new chemical complex in Port Harcourt won him the Nigerian Journalist of the Year award in 1987.

A year later he had the scoop of a lifetime, a news story that hit the headlines worldwide. After six months' investigation of the trade in toxic waste in Europe and Nigeria, Ogunseitan unravelled what became known as the Koko affair. He discovered that at least twenty extremely poisonous chemicals from a number of European firms had been dumped in Koko, a quiet town in the swampy delta of the Niger river, about 240 kilometres east of Lagos. Thousands of drums, bags and containers, many of them broken, were found in an unguarded compound, surrounded by only a ten-foot high, rickety fence. Less than fifty metres away people were living quietly, cropping cassava.

The Koko affair shocked Nigeria, and fuelled outrage throughout the continent. Developed nations were using Africa as a dumping ground for toxic waste. The Nigerian Government had just led a high profile, international campaign to persuade other African nations to unite against toxic waste dumping. At a meeting of the Organization of African Unity, the Nigerian President Babangida had declared only a month before that 'no government, no matter what the financial inducement, has the right to mortgage the destiny of future generations of African children' by allowing imports of toxic waste: 'A crime against Africa'. Now tons of European industrial waste, mostly the highly toxic PCBs, were found in his own backyard.

The reaction of many Nigerians was indignation, Ogunseitan recalls, especially because the waste came from abroad. "People didn't realize that we also have a lot of locally produced waste, including PCBs, because adequate information is not available about that. People working in ports and factories, for example, should know which chemicals are dangerous to themselves and to the environment, so they can inform their unions or the press. Information is vital. Even scientists and journalists don't have reliable and up-to-date facts."

Late in 1988 Ogunseitan visited the United States, where he talked about the dumping of toxic waste. "While there, it dawned on me that we do not have enough well-informed reporters. So I said to myself: Why not set up a specialized information agency? A sort of databank to gather and disseminate information for scientists, journalists and others. Not only about the environment, but also about health, population and development issues, science and technology."

Better access to all types of information is essential for the country's development, Ogunseitan felt. An information agency was needed to provide scientists, students, reporters and others with the information they need. Ogunseitan quit his job on 'The Guardian' to establish the African Centre for Science and Development Information (ACSDI).

Crisis in science education

Back in Nigeria, Ogunseitan began to look for confirmation of his observations on the need for such an information agency. In a newspaper article he offered to make available copies of fifty papers to be presented at the 1989 Scientific Conference of the American Association for the Advancement of Science (AAAS), in Washington, D.C. From the response he hoped to get an indication of the need for an information dissemination system.

Six weeks after the offer was published, 129 requests had been received for copies of the papers from reputed scientists, professors, lecturers, postgraduates and undergraduate students across the country. This response reinforced the need for Ogunseitan for his information centre and confirmed earlier observations.

In 1988 he had made a study on the state of scientific education in Nigeria, for which he had interviewed more than fifty teachers and researchers in eighteen institutions of higher learning across the country, and forty-three students. The study resulted in a special newspaper supplement titled 'Crisis in Science Education', which painted a bleak picture of the state of academic life in Nigeria.

In this publication Ogunseitan quotes a retiring professor from the University of Ibadan as saying: "When you call some of us scientists, we laugh at ourselves. We know we can no longer make contributions to science. I do not know what my colleagues in Kenya or London have found, for example. So I can not carry out an experiment and believe I am on the path to an original contribution to the sciences. If I have been giving generations of students the same notes for the last ten years, I should not call myself a scientist." A young physics lecturer from the University of Benin is quoted as saying, "There is no more science in what we are doing."

Ogunseitan propounds: "Many people in our universities are not sure what is the state of science. Scientists often have to rely on what they are told, for example by newspapers, by friends or by 'Time' magazine. But if that is what you have to rely on, you are never going to be an authoritative and confident scientist."

Oil-rich Nigeria had a booming economy in the seventies. But with the drop in the price of crude oil on the world market in 1986, foreign currency earnings fell dramatically. A deep crisis in the heavily indebted, import oriented economy was unavoidable. The military government introduced belt-tightening measures, which not only grossly reduced the income of researchers, but also led to severe cuts in funds available to their institutions.

"As with many other national institutions, the scientific community has been marginalized economically," Ogunseitan observes. "Its members can no longer afford to subscribe to scientific journals and professional bulletins published outside the country. Scientists can no longer pay their way to conferences outside Nigeria, and even the national associations of scientists find it difficult to organize regular meetings at local level."

"Currently, the best of Nigeria's scientific manpower is being drained, mainly to the Middle East. Although the government is trying to check this brain drain, scientists are finding abroad much that is lacking in Nigerian institutions."

Ogunseitan is convinced that the serious infrastructural deficiencies strangling science in Nigeria today can be and have to be
overcome, not only for the good of Nigeria but for the good of the
world.

Information to bridge the scientific gap

"An uninformed Africa is as much a threat to Europe, the Americas
and Japan as it is to Africa and Africans themselves. Access to
information is vital for every country, but there is an enormous
imbalance in a world in which one part lacks even the most basic
information."

"Like the Brazilians, Nigerians need to know more about why
the world wants to reduce logging activities in the rain forests. That
is an information problem. A Nigerian professor of environmental
sciences is in a better position to explain to a Nigerian audience the
dangers of unchecked destruction of the tropical forests than an
American, even if the latter is the world's leading authority on the
dynamics of global climate, the ozone layer problem and the
greenhouse effect. The lack of adequate and effective information
flows to developing countries has made an holistic approach to
essentially global problems impossible."

"I am trying to stress the importance of access to up-to-date
information. If science is to benefit society as a whole, researchers in
developing countries must have access to information and to the
latest developments in science. To do otherwise is to take the
fatalistic view that people in the developing world are not part of
the globe."

"But when you look at the drastic economic restructuring that
takes place in these countries, and the unfair contest for government
attention between education and other sectors of the economy, you
will realize that the conventional system, whereby researchers pay
subscription fees and receive their journals regularly, can not be
sustained."

Ogunseitan decided to do something about the problem and
developed his idea of an information agency. He wrote an am-
bitious proposal for, as he formulated, "setting up an institution to

arrest the decline of science in Nigeria and to restore and strengthen science support systems in Africa."

The centre should link Nigerian scientists with their colleagues in Europe, Asia and the Americas by facilitating access to an exhaustive number of scientific journals and other academic publications. It should establish and maintain a science and development information databank and pool for Nigerian researchers and others who are interested in sub-Saharan Africa.

"Not just a one-way flow of information," he explains. "People in the United States doing research on malaria, for example, may well like to know what is being done here by informed researchers who live with the disease, see patients every day and who don't have to simulate conditions. Why not collaborate? Otherwise you may be wasting money."

The centre should furthermore create a framework to link scientists in Nigeria and later scientists across Africa for the purpose of exchanging scientific information. It should strengthen local scientific bodies, provide back-up services for science desks in media institutions, as well as stimulate the development of a corps of information gatherers and reporters on science and development.

"Strong local science reporting is desirable, not only to de-mystify and create awareness of the benefits of science to a developing economy, but also to entrench science in the thinking of the people."

Information for Africa

In 1990 the African Centre for Science and Development Information is, to use Ogunseitan's own words, "in the latest stage of its formative years." It has been incorporated under Nigerian company registration law, and is functioning on a small scale, providing 'skeleton services'.

"I have converted eighty percent of my apartment into an office," Ogunseitan explains in his living-and-working quarters in Palmgrove, Lagos. The rooms are filled with huge piles of newspapers, magazines and journals, filing cabinets, bookcases, a typewriter and two personal computers. An oil lamp illuminates the room until the latest power cut is over. A donation from CUSO (formerly Canadian University Services Overseas) has provided a quarter of the start-up funding. Private donors and the founder himself are taking care of the rest. "We now have three people: a development journalist, a typist and me. We will be joined shortly by two science graduates."

ACSDI has the beginnings of a documentation system, and contacts with a great number of international development organizations, within and outside Nigeria, are being established to gain support for various activities. The centre provides 'The Guardian' with an article every week. It unveiled a toxic meat scandal and brought forward scientific information in a highly publicized affair of poisonous cassava.

"We are now seeking the best information system for our purpose. A directory of resource persons around the country is being developed to be available to everyone. We are thinking of producing a rural information newsletter and an environment and development magazine. We fill a full page in 'The Guardian' every week and we intend to deliver the same service to at least five other national daily newspapers."

But the ambitions go much further. To fulfil its role as a scientific information agency, the centre will have to subscribe to an exhaustive number of scientific journals and magazines. Ogunseitan's objective is to obtain rights to reproduce articles by photocopying, and to distribute copies to researchers and scientists. "The interested scientist will only pay a minimal charge to cover costs. All subscribers will receive free of charge copies of abstracts of all the articles in their particular disciplines."

Asked whether it is not naïve to suppose the centre will be able to obtain copyright material without paying the full amount, Ogunseitan answers: "Information flow is highly commercialized today, particularly in the West. But that is exactly why only a special arrangement can direct the information flow to the developing world. Only then can science be made meaningful to these societies."

Ogunseitan is convinced that the agency should always remain independent of any government, although he acknowledges it will need the support of grants and donations of national and international organizations and agencies. While ACSDI is a non-profit organization, the ultimate goal is to be self-supporting. Ogunseitan has no illusions about the financial support needed in the initial years: 400,000 US dollars is the budget estimate for the next two years. Ogunseitan fully realizes how much has to be done before the full-scale operations of ACSDI can start. Organizational and other types of support from at home and abroad are essential.

Juurd Eijsvoogel

The developed countries are building integrated information systems, while in Nigeria the oil lamp is still serving during the power cuts. But financial assistance for Ogunseitan's ACSDI is hardly to be expected from Federico Mayor Zaragoza, Director General of Unesco, the United Nations Educational, Scientific and Cultural Organization. His organization could play an essential role in bringing scientists of East and West, and North and South together. One of Unesco's missions is to push scientific research and development ahead in the Third World. "While it is often stated that Unesco's budget is smaller than that of a good university, we do not work with money alone. We work with ideas, initiatives and projects. A good idea has the power of transformation. Change is produced by new ideas, not by money," says Mayor.

FEDERICO MAYOR ZARAGOZA, the diplomat

Coordination without funds

Federico Mayor is at ease in many different roles. He is a scientist, a politician, a writer, a manager of a large bureaucracy and a top diplomat in the world of the United Nations. "Can we afford to be complacent about the one billion or more people in the world today who are illiterate?," Mayor questions, pointing at the growing information gap between North and South. To draw attention to this problem Unesco has proclaimed 1990 the International Literacy Year. According to the organization, in 1985 one in every four adults was illiterate. In the Unesco definition, 'these people can not with understanding both read and write a short simple statement on everyday life'. Ninety-five percent of them live in the developing world, seventy-five percent in nine countries: India, China, Pakistan, Bangladesh, Indonesia, Nigeria, Egypt, Iran and Brazil. Mayor is concerned that the focus of the developed world on East-West relations will overshadow North-South relations. "The eighties were extremely hard times for the poor South countries, especially the less developed of them."

Science cooperation

Unesco's mission goes beyond stimulating literacy programmes, affirms Mayor. Scientific programmes should be pushed as well. "We are trying to persuade all these developing countries that they should have a corps of scientists, be it a modest one, to select and adapt knowledge to repair foreign technology, otherwise there is only competition among the scientists of the most industrialized countries. For poor countries too it is essential to have scientists."

Mayor says he bases this opinion on his own personal experience in Spain, where he was born in Barcelona in 1934. In Madrid, the young Mayor, son of an executive in the pharmaceutical industry, studied pharmacy. He became a research scientist and in 1963 was appointed Professor of Biochemistry at the University of Granada.

He bridged the gap between science and society early in his career and finally crossed the bridge, although he still considers himself to be a scientist: "As a young scientist in a very old university dating from 1517, I soon realized that this centre of learning, where so many brilliant people had worked, has no impact whatsoever on the society around it," he says. "The university existed in a world apart. There and then I decided that wisdom is the application of science for the benefit of society. I became a politician."

At the age of 33, Federico Mayor became Rector of the University of Granada. Under the Franco dictatorship this automatically gave him a seat in Parliament. For some time he was Secretary of Education in the Franco Government. Later, in the democratically elected Government of Adolfo Suarez, he was Minister of Education and Science from 1981 to 1982. At Unesco headquarters in Paris, Mayor was Deputy Director General from 1978 to 1981, and later returned for a year as a Special Advisor to the controversial Senegalese Director General, Amadou-Mahtar M'Bow.

Mayor was a member of the European Parliament when an international group of scientists proposed him as M'Bow's successor. In 1987 he was elected to the top of the troubled Unesco, just after the loss of two of its founding members and major financial contributors, the United States of America and the United Kingdom.

"When I was Professor of Biochemistry in Granada, I found myself in the situation in which science was not supported. Gradually, in the sixties, and gathering momentum in the following decades, science gained Government support in Spain. Now 0.3 to 0.4 percent of the Gross National Product (GNP) is devoted to science. In developed countries this figure is at least 2.5 or 2.6 percent. In the last year, it was 2.8 percent in the United States and even 3.0 percent in the Federal Republic of Germany."

"The greatest disparity today between industrialized and developing countries is in investment in research and development. Some developing countries are investing quite heavily in education, but when it comes to science, there is a big gap between them and the developed world."

"Many developing countries consider it easier to buy technology from abroad. But I keep repeating that research, whether done by national or foreign scientists, has to be paid for in one way or another. Applied science needs basic science to be applied. My response to ministers from developing countries who proclaim that applied science must be strengthened, is that applied science does not exist, only science that can be applied. Therefore you must contribute, even if only very modestly, to the development of science."

"I think we must aim for solidarity. Some centres of research in the North have twin agreements with centres in the South. Various associations of scientists in the developed world could establish exchange agreements, and universities could set up cooperative arrangements giving access to bibliographies, equipment, training courses, and the like."

Science in decision making

"Unesco must be a bridge. It must provide the means, on the one hand, for people in developing countries to exchange views, to obtain fellowships and scholarships, to participate and assist in training courses, to do research in laboratories in developed countries. On the other hand, Unesco must stimulate awareness in scientists from industrialized countries of their colleagues working under less favourable conditions than themselves."

"Now that East-West relations have improved, it is perhaps appropriate to recall, with due respect, Unesco's role in this process. Over the years the organization has actively stimulated contact between the East and West scientific circles, including fellowships and scholarships, meetings and workshops, with the emphasis on networks and contacts."

Science, according to Mayor, will play an increasingly important role in politics all over the world. "Every day we see a further 'scientification' in decision making, and I am convinced this process will continue. Politicians are more and more looking to the scientific

community for help in solving problems. I have the opportunity
and the honour of meeting Heads of State, Presidents, Ministers of
Economic Planning and Education. You can not image how they
are constantly confronted with problems of the environment, genetic
manipulation etc. They need information from the scientific community."

"This means, of course, that scientists need to understand the
problems and challenges of society at large as well as of individual
groups. The opposite is also the case, and that is what I'm more
interested in as Director General of Unesco. We must try to make
the views of scientists available to the decision makers. Not only do
politicians want to know what scientists are doing, they have to
know."

"What should be done about the environment, for example?
Many responsible world leaders are alarmed about the situation.
They are aware of the greenhouse effect, ocean pollution and
excessive use of fertilizers and want to take action. However, they
need information from the scientists on which to base their decisions."

"Today, this is one of the most important roles of Unesco.
Professor M.G.K. Menon, the President of ICSU, the International
Council of Scientific Unions, stated recently that it is essential that
scientists address governments. Most forums are government-to-
government, but governments must also talk to the scientists who in
their turn must come down from their ivory towers. They must be
aware of what is happening in the real world, in public life, and
enlighten the world with their vision."

The S of Science

"Unesco has the S of science, and we must make the bridge with
the scientific community," Mayor affirms. "This is essential for the
future of mankind, I am totally convinced. Therefore we are going
to underline the S of Unesco in the years to come."

"To come back to environment," he continues, "we need a
channel for the communication of information. At present, scientific
information on a particular subject tends to be dispersed under the
headings of geology, biology and marine ecology, for example.

Scientific research institutions are also dispersed. In the United Nations systems, several institutions deal with environmental issues but are coordinated by UNEP, the United Nations Environmental Program. FAO, the Food and Agricultural Organization, deals with soils, afforestation and deforestation. WMO, the World Meteorological Organization, is concerned with meteorological changes, while Unesco has programmes on geology, oceans and biosphere, and the convention of the natural heritage."

"These activities need to be coordinated and integrated, and not only in the United Nations system. We must work together with the intergovernmental organizations: the European Commission, the Organization of African Unity, the Association of South-East Asian Nations, to name a few."

"We must also have the scientists at our side, otherwise we are still talking government to government. We need their insight and the fresh breath of scientific inspiration. I am glad that the Second World Climate Conference in 1990 is being organized by UNEP, WMO, Unesco, and particularly ICSU, which represents the scientists. This is the first time ever, I think, that the scientific community is a partner in organizing such a meeting."

"I have suggested to the other organizations, including scientists, that we should present a concise but comprehensive statement on the state of the environment to Heads of States every year. This statement should set out priorities and measures as well as significant trends in ecological evolution. This will not be as easy as it sounds. It is difficult to get agreement, but nevertheless global challenges cannot be faced alone, not even if you are a giant."

In his sixth floor office in the Unesco building, overlooking the stately, sand-coloured buildings of central Paris, Federico Mayor shows a fervour and a sense of urgency not often found among bureaucrats. "We must not delay treatment of some aspects of ecological conditions, because tomorrow it may be too late," he asserts.

"I am a biochemist and for 25 years I have been working on the human brain of newly-born infants. I understand the urgency of early and timely diagnosis. At the time of birth already, any diagnosis of a patient's condition, no matter how well considered, is of little use if the patient dies or if the brain is irreparably injured before the diagnosis can be pronounced. Therefore, we must act now. And that is the only right diagnosis. Timely diagnosis of certain aspects of the environment is essential so that treatment can begin before the point of no return is reached."

Utopia

Mayor says that Unesco is trying to contribute to international dialogue and cooperation. "The result must be peace, because, to quote our preambula, as war starts in the minds of men, we must build the defences of peace in the minds of men. Unesco's mission is 'to contribute through education, science, culture and communication to sustainable peace'."

Are these realistic goals for a large bureaucratic international organization such as Unesco? Aren't these grand words a little naïve? "While in global terms we may be very small, we have immense power and strength, precisely because of our ideals. We have the power of knowledge, of the millions of teachers throughout the world," says Mayor, "of research scientists, artists and journalists. The greatest power is the intellect to persuade and to convince. Even without money, if we are able to persuade a nation that education should be the first national priority, then by word alone we shall have launched an immense action in that country."

"Unesco is more than a technical agency: it is an ideal. As with all ideals, it seems to be a Utopia. Yet so many of yesterday's Utopias are today's realities. I like the saying that 'only those able to see the invisible are able to do the impossible'. We must have this kind of passion in order to be able to transform today's ideals into tomorrow's reality."

The New Page

In 1987, Federico Mayor, author of numerous scientific publications and several collections of poetry, published an essay on the world's problems called 'Tomorrow Is Always Late'. His latest book is more optimistically entitled 'The New Page'.

"I like to think we are beginning to write a new page of a new era in civilization: the peace culture," he muses. "From the beginning of history, we have been living in a culture of war. We have had to defend our territory from invasion. History has, in fact, been a description of winners and losers."

"Since the forties, however, we have come to realize that we are all losers in war and that therefore war makes no sense. We are starting a new age that will be written in a new language in which intercultural dialogue will be possible. Now we have to find out which is this language. I think this language will encompass the promotion of creativity, enjoyment of leisure, access to knowledge and learning for everybody, thus having criteria for welfare other than the economic growth, such as human rights, ecological and ethical values; criteria that have been put aside too much by the economocentric vision of the world."

"All this is 'The New Page', which contains a feeling of hope. There has been a tremendous transition within a very short period of time from a situation of confrontation between the two super-powers to the conviction that in a war everyone is a loser. We are entering into a new era that coincides with a new millennium."

Juurd Eijsvoogel

"I understand the urgency of early and timely diagnosis," says Federico Mayor. So does Dr. Alexander King, President and one of the founders of the Club of Rome, which almost twenty years ago shocked the world with the pessimistic report 'The Limits to Growth: a report for the Club of Rome's Project on the Predicament of Mankind'. Since the publication of the report, King's mission has lost nothing of its urgency. King is still travelling the world speaking with undiminished fervour about the global problems facing humanity. Population growth, industrialization, environmental pollution, food shortages and lack of long term thinking are as worrying as ever. Even the depletion of natural resources, which turned out to be less alarming than predicted, gives no grounds for complacency, King warns.

ALEXANDER KING, the activist

The human lemmings

Alexander King does not speak as a scientist, although he had a distinguished scientific career at the Imperial College in London; nor as a politician, although his mission often leads him into political circles; nor as a representative of an international organization, even though he was Director General for Scientific Affairs of the Organization for Economic Cooperation and Development (OECD) and Chairman of the International Federation of Institutes of Advanced Studies. The President of the Club of Rome speaks as a concerned individual, trying to influence policy makers and public opinion.

The Club of Rome is, in King's words, "a loose association of some 100 individuals from forty or so countries on five continents. They represent diverse ideologies and occupations, and endeavour to influence decision makers and the public by identifying and analysing global issues."

Being completely independent and free from formal commitment is essential to King's mission. "Constrained by their bureaucracy and the tyranny of the next election, politicians and governments are not able to look very far beyond the ends of their noses," he declares, "whereas long-term thinking is one of the fundamental aims of the Club of Rome since its foundation. It is one of the three criteria for our work. The other two criteria are the need for global thinking, because we foresaw many problems that couldn't be tackled by individual countries alone, and the interaction of these problems, what we called the *world problématique*. The main problems of contemporary society are like an untidy ball of strings: by pulling out one thread all the others begin to come with it."

The Club of Rome

"We planned the Club of Rome as a club, and not an organization, because we wanted no bureaucracy. In both appearance and reality it had to be independent. We were not going to take money from any government or enterprise. We worked for twenty years without any budget, and no one has ever been paid. As far as I am concerned, it is the most demanding job I have ever had: unpaid and very expensive. The Secretary-General Bernhard Schneider gave sixty percent of his time for free; we managed to keep it very flexible."

The Club of Rome has many scientists among its members, but never uses their expertise to present unanimous, scientifically-based recommendations to the world. "Our reports are not made by the Club, but to the Club. We want to avoid having to go through the process of reaching a concensus. As our members have widely varying backgrounds, ranging from American bankers to Soviet marxists, we can't expect them to agree on everything. We select problems and ask someone, or a group, to write a report to us. If we had to reach a concensus, it would mean too much of a compromise."

'The Limits to Growth' was outspoken enough to inspire frantic discussion as well as sharp criticism. The rather apocalyptic message was that if growth in world population, industrialization, pollution, food shortage and depletion of natural resources continued unchecked, the limits of growth would be reached within a hundred years. Aurelio Peccei, then President of the Club, wrote in a later book that 'the human race is hurtling towards disaster.'

"It created a scandal and immediate interest," King reflects on the book's reception. "By now it has probably sold about twelve million copies in about thirty languages, which is not bad for a non-fictional work. While far from being perfect, it was a pioneer work which has been improved on since then. Most probably the book gave rise to debate throughout the world. And that was very badly needed."

"In one way the book made our name, in another way it hurt us. 'The Limits to Growth' was seen by many people of the South as an attempt by imperialist countries of the North to tell the South to stop making so many children so that we, who have fouled up the world and made all the mess of everything, could go on as we are."

"And in the North most classical economists were utterly hostile. Only a few were not. Many people identified us with zero economic growth. On the basis of our work, the Chairman of the European Commission, the Dutchman Sicco Mansholt, wrote to all European governments advocating zero growth. That was not our intention! We wished to question the wisdom of purely quantitative economic growth and to give a more holistic view of what it was leading to if policies were not changed."

Policies have not changed greatly since then, King has to admit. Nevertheless, he does not think that the Club's mission has failed. "There is a far greater awareness of the problems facing humanity. Undoubtedly there is more public understanding now. We didn't think it would go as quickly as this. Contributing to this level of awareness has been one of the greatest achievements of the Club. Governments, political bodies, business and labour organizations, environmentalists, academics, religious groups and concerned people everywhere are all grappling with the same set of problems. In the scientific community too there is now a sense of urgency."

"Indirectly, 'The Limits to Growth' has influenced policy makers," King considers. "I was told by a senior Canadian Minister, for instance, that never again would we be able to think of individual sectors of the *world problématique* separately. The Club of Rome was the first to talk of global problems, and to point out the environmental problem.

Environmental questions

In 1987, the World Commission on Environment and Development, established by the United Nations General Assembly and headed by the then Prime Minister of Norway, Gro Harlem Brundtland, presented the famous report, 'Our Common Future'. Just like the Club of Rome, the Brundtland Commission was convinced of the need for a global and longer term perspective, and to consider none of the major problems in isolation.

"When the terms of reference of our Commission were originally discussed in 1982," the President writes in her foreword, "there were those who wanted its considerations to be limited to 'environmental issues' only. This would have been a grave mistake. The environment does not exist as a sphere separate from human actions, ambitions and needs, and attempts to defend it in isolation from human concerns have given the very word 'environment' a connotation of naïveté in some political circles. The word 'development' has also been narrowed by some into a very limited focus, along the lines of 'what poor countries should do to become richer'."

In spite of the adherence of the Brundtland Commission to the principle of holistic thinking and the linkage of political, economic, social, technological and environmental problems, Alexander King is not enthusiastic about 'Our Common Future'. He doubts whether sustainable economic growth can be environmentally acceptable in our present economic system based on more and more consumption. "It is a compromise," he feels. "It is not brave enough."

"The report is extremely useful," he continues, "however, the members of the Brundtland Commission are very near to the decision-making process in their countries, and so could not be as independent as we are."

Alexander King has been called 'The Original Green', but he does not feel akin to Green political parties. "We are in a transitional period now. Environmental concerns are no longer for the Greens only, but have become major elements of economic, industrial and agricultural policy. The Greens are too unisectoral," suggests King. And, as a scientist, a civil servant and an activist for a better future, King has always been in revolt against unilateral approaches to problems.

Multidisciplinary research

As a student of chemistry at the Royal College of Science in London, King felt little at ease with the dominant 'narrow concept of science', as he calls it. Science, according to King, should be an integral part of contemporary culture, particularly in Britain and America. "In the English language 'science' means natural science, where as in French, or even in Russian, it means knowledge in a broad sense," he explains.

King's career, from scientist to activist, illustrates his gradual awakening to the wider problems of humanity. Born in 1909 in Glasgow, he went to English public schools and obtained his doctorate in chemistry at the Imperial College in London, where he was later a lecturer until 1940. During the Second World War, King worked for the British Government. "I was not a scientist then, but a scientific advisor. At first I hated the idea of bureaucracy, because as all scientists, I was conditioned to appreciate the freedom of research. At the age of thirty-two, I became Deputy Advisor to the Minister of Production and very shortly found myself in charge of a crisis situation, namely the urgent need for an effective insecticide."

By coincidence, King played an important part in the application of a now notorious chemical. "A letter, intercepted by the wartime censor, from Geigy in Basel to Geigy in Manchester, passed over my desk. It stated that an excellent chemical agent had been found to protect ladies' furs: dichloro-diphenyl-trichloro-ethane. I thought, if it is safe enough for women's furs, then it's worth testing. It turned out to be the best insecticide ever known. We called it DDT. Within three months, it was operational on the Western front, and at the end of the war, I received a letter from Lord Mountbatten declaring that this chemical had saved him three-quarters of a million casualties."

"So you see, being responsible for this, makes me a very wicked person in the eyes of the Greens. Nevertheless, DDT saved enormous numbers of lives and also prevented illness in many more people." Later the use of DDT was widely condemned on the grounds of the harmful toxic effects on plants and animals, including human beings. Alexander King reflects: "My own doubts came when DDT was introduced for civilian use. In Guyana, within two years it had almost eliminated malaria, but at the same time the birth rate had doubled. So my chief quarrel with DDT in hindsight is that it has greatly added to the population problem. Of course, I can't play God on that one," he muses.

When the war came to an end, King had been away from research for almost five years, "at an age which should have been the optimum for creativity. I wasn't sure I wanted to go back to scientific research." King continued his career in the British civil service and was eventually nominated Chief Scientist of the Department of Scientific and Industrial Research. He became interested in multidisciplinary research and became convinced of the need to, as he puts it, "bring in the social sciences."

"Engineers, scientists, social scientists and economists would work together on problems. In 1957 we completed a study of automation. This was somewhat premature because micro-electronics had not been invented at that stage. Nevertheless, I had a team comprising an electronics engineer, a mechanical engineer, an economist, an industrial sociologist and others. Our report happened to coincide with a strike, called an automation strike, and the government of the day was surprised and very happy that they had a full report on automation."

"It was a bestseller for our department, but they didn't like it. Because it was neither chemistry nor physics, and certainly not science as they saw it in their ivory tower approach."

Make governments jump

King left the service of the British Government and went to Paris as Co-Director of the then European Productivity Agency, EPA. In 1961 he came to the newly founded Organization for Economic Cooperation and Development as Director for Scientific Affairs, and later became its Director General for Science, Technology and Education.

"At that time the concept of science policy was being developed. We began to look at science and establish priorities, which was still controversial. Even at the ministerial level some people regarded the discussion of science policy in an economic context as a kind of prostitution."

"By 1967 we were very worried about excessive interest in economic growth, almost for its own sake. We felt there were many signs of unfortunate by-products and side-effects. And in 1968 as a result of these concerns I decided with the agreement of the Head of the OECD, to try to create an independent group to look at the long-term issues of Europe without political bias and government responsibility. We looked to this group for constructive criticism of government action and thinking, and every now and then to stick pins into governments to make them jump."

"Then by a series of accidents I met the Italian industrialist Aurelio Peccei, and we called a meeting in the Academy of Sciences in Rome, to which some thirty personalities were invited to discuss long-term European problems. The meeting was a complete failure. The French argued semantics; the Italians were obsessed by the Vietnam war; and there was a general anti-American atmosphere. It ended without any results."

"That night, six of us dining at Aurelio Peccei's apartment, agreed that we had been too naïve in calling this meeting. There and then, we decided to form a small club, really for our own mutual education. For obvious reasons, we called it the Club of Rome, but later this led to many misunderstandings. Some people thought we were an emanation of the Treaty of Rome; some thought we must be a hidden branch of the Vatican; and others that we must be fascists in the image of ancient imperial Rome."

"For about a year and a half, the original members met every six weeks or so, generally in Geneva. Aurelio Peccei was the President, and I took over when he died in 1984. Gradually more people were brought in, and in 1970, a very lively discussion led to the suggestion that we should undertake a computerized study of the *problématique*. We wanted to determine the physical limits to population growth and man's material activity."

"Our search for an academic group with the techniques and methodology to do this study ended with J.W. Forrester, at the Massachusetts Institute of Technology, in the USA. This led in 1972 to 'The Limits to Growth', about the nature and consequences of exponential growth. Since then about eighteen reports have been published including 'No Limits to Learning', on the need to change attitudes about educational systems. 'Restructuring the International Order' and 'Micro-Electronics and Society: for better or for worse', on the consequences of computerization of the world, are two major reports."

"More recently, a very successful book was 'The Barefoot Revolution', a study of spontaneous efforts by groups of peasants and villagers throughout the underdeveloped world, to help themselves, in despair of getting help from the government and others. 'Africa, Beyond Drought and Famine' is another recent report."

Alexander King is more a man of action than of reflection. But his action aims to provoke discussion and reflection on a world scale, resulting, he hopes, in beneficial change in man's behaviour. "The meeting in 1989, 'Global industrialization: panacea or nightmare?' discussed the greenhouse effect. Can economic growth in a purely material sense continue? Can the consumer-driven society persist? What is going to happen to the less-developed countries? One of the suggestions to the United Nations was to set up a security council for the environment."

"I regard the increase in the totality of human activity as the basic menace. This is partly due to the number of people, and partly to the amount consumed of energy, raw materials, goods and services. When I was born the world population was 1.8 billion; it is now over five billion, and will approach six billion at the turn of the century. Per capita consumption has increased enormously. At the time of the industrial revolution, 200 years ago, the average European country had about the same level of consumption as the average underdeveloped country today."

"The only approach to the *problématique* is to encourage solidarity between people, based on common self-interest. We are all selfish, we all have self-interest. But with all these world threats, there is a common self-interest to resist annihilation of our race, and to provide a world decent enough for our children and grandchildren. Most people have some interest in this, not all, but most have. And that is encouraging."

Juurd Eijsvoogel

Alexander King, the activist, stresses the importance of international solidarity, because, he says, with problems being more and more global, there is a common self-interest to resist annihilation of our race. Erich Bloch, the policy maker for US science, does not use such dramatic words, but he shares the notion that growing international cooperation is of the utmost importance. "Scientists of all countries, from the North and South, the East and West should work together, because the world is confronted with problems that cross boundaries, such as conservation of the global environment," he says.

ERICH BLOCH, the science policy-maker

The government catalysis

Bloch, who has a reputation for toughness and abrasiveness, is the powerful and controversial Director of America's National Science Foundation (NSF). In his spacious office not far from the White House, he points out that science plays a somewhat ambivalent part. "One of my concerns is the damaging effect that science can have on the environment," he says. "The application of science has undoubtedly contributed to the problem. But although science is part of the problem, it is also part of the solution in that scientists are attempting to understand what is going on, trying to find new non-polluting materials. Those materials don't appear overnight from nowhere, and we can not go back in time and restore the environment to its former pristine self. We must go forward with new ideas and developments, new inventions and innovations, which do not have such polluting effects or damaging characteristics. There are two sides to the coin, science has always had those two sides."

In his urge to go forward Bloch persuaded US Congress to increase his agency's budget to an unprecedented US $1.8 billion. And he insists that further funding boosts are needed, because of the increasing importance of science and technology. In 1985, he was awarded the National Medal of Technology by President Reagan, for his contribution to pioneering developments related to the IBM/360 computer that revolutionized the computer industry.

Bloch was born in 1925 in Sulzberg, Germany. Before joining the NSF, Bloch spent three years as Chairman of the Semi-Conductor Research Cooperative, a group of leading computer and electronics firms that fund advanced research in universities and share the results. Prior to 1981, he was a Corporate Vice President for

Technical Personnel Development at IBM where he had worked since 1952, originally as an electrical engineer.

Since his appointment was confirmed by the US Senate in 1984, Bloch has expanded the NSF's mission to support basic scientific research by adding technology and collaboration with industry to the agency's areas of interest. He is respected by the US Administration and by Congress and is considered to be one of the main individuals who will invigorate policy-making in science and technology.

Educating society

"The impact of science on society is obviously very profound," he says. "We are going through a complete restructuring of our society right now, in as much as each country and every economy will be much more dependent on knowledge than previously. They used to be highly dependent on natural resources. Now, the economic competitiveness of a country will depend more on human resources, on people."

"It is evident today in the new industries that are springing up, such as biotechnology and semi-conductor firms, in the new synthetic materials being used, and so on. What a job demands in terms of skills and background and what you need to know nowadays, is also completely different from what it was in the past. I think this change will accelerate in the next fifty years, and maybe beyond. That's what science and engineering can do to society, turn it upside down. We live in a completely different world than the one that existed before World War II."

"In this highly technical world, it's not easy to attain good insight into trends in scientific impacts on society," Bloch goes on. "It is extremely difficult to stay up-to-date, especially for politicians and governments, the people who have to make the decisions. They don't really have the necessary background for it. And it's equally difficult for the public to know what's going on as we've done a poor job educating them. We should focus more on education, not because people want to become scientists, but so that they can understand scientific principles and can form their own opinions."

One of Bloch's major concerns is that society should not become divided by education because some people are technically skilled and some are not. He feels there is a danger that those who are technically illiterate may not be able to form their own opinions, and will no longer play a part in the democratic process because they do not understand all the issues. They may simply regard science as the only way to solve problems. "If science is seen as a sort of almighty God who rules the world, then that's a problem," Bloch explains. "Science is not almighty. I'd rather look at it in a different way. Too often, we fail to combine scientific knowledge with sociological understanding. We look at a problem with a scientific and not a sociological eye, and vice versa. We have to get to the point where we tackle problems in an interdisciplinary way."

"How do we do this? Well it's not easy, otherwise it would obviously have been done. I don't know quite how you achieve it. By bringing people together, perhaps, people that represent various disciplines and different understandings about a particular issue, instead of solving problems in a peacemeal fashion and in isolation. There is a trend in that direction, maybe not very pronounced yet, but I encourage it very strongly."

"It's not only education of the general public which is at stake," Bloch goes on to say. "In the United States we are even doing a poor job educating our students. We need to focus more on science and mathematics in high schools and colleges, because we are not in a very good position. The NSF foresees a US shortfall of 560,000 scientists and engineers in the year 2010. And many graduate departments for science and engineering are entirely populated by foreign students. We can't seem to function in this country without this tremendous input of foreign students."

Bloch points out that scientists and engineers nowadays experience special problems because of the rapid changes. "Today, the occupational approach is as important as experimentation and theory were at one time. The modeling, the visualization of inventions – that's a different approach to science. Many of our scientists are not accustomed to that, they weren't brought up that way."

"How should we address this problem? By educating from the base, by focusing more on these issues in the popular press and in the political environment. How else can you draw attention to it! The problem is that the media, at least here in the United States, are not focusing enough on it. Similar changes have obviously occurred in the past, but I think that those taking place today are much more profound. They are also happening much more rapidly and involve the whole world. The Industrial Revolution in England encompassed perhaps one eighth of the world, I have not figured it out precisely. But this knowledge revolution really affects the whole universe, and that's different."

"The mere fact that more countries and more people are involved in this type of revolution, accelerates the process. There is now greater international cooperation and there should be even more. It is never enough. A big effort is being made in this respect, for example, in Europe, the United States, Japan and several other countries. A new kind of scientific cooperation will hopefully develop now that Eastern Europe is opening up. A certain amount of such cooperation even emerged with China, before the sliding back that is, before the events of Tiananmen Square."

The role of government

Bloch believes that the complexity of the world calls for increasing interaction between governments, industry and the public. He warns against imbalances in the system, where industry or government have the upper hand, although he agrees that government should sometimes control scientific research for ethical reasons. "Government control may hamper innovation," states Bloch, "and may also direct science into one particular area, whereas it should be broader." He gives an example of a situation within the Bush Administration where funds have been withheld for experiments involving fetal tissue, although such research might find a cure for Alzheimer Disease. He also cites Allan Bromley, Science Adviser to President Bush, who said in an interview that the US government might reach the point where it has to stop funding life-extending technology. Bromley considers that this will be one of the greatest crises of the next decade, particularly as there is no fully developed valuation system for determining how withholding technology can be done ethically.

"But you have to make a distinction between real concerns about excessive government control and ideological concerns," Bloch continues. "If the concerns are ideological, then there is a serious problem which can only be solved by politicians and by people through the ballot box. Basic science and technology must find a common cause with social science, humanists and religion. As I stressed before, people from different disciplines must be brought together."

"But there can never be complete privatization of science and technology. Government has to be involved because it represents the common people and works for the common cause. Government has to look at science as it looks at the infrastructure of a country so that it is responsible for defence, health, the environment, education, as well as science. It should all be seen as part of the same process. It's debatable to what extent a government ought to be involved with science as this depends on many things, such as the culture, history and tradition of a country, and the way the governing system has been set up. This is different elsewhere, but I don't know of one modern country that doesn't think it has a responsibility for research and education."

With its free market ideology, fear of extensive bureaucracy and instinctive aversion of government control, the United States should develop a different attitude to government funding, Bloch argues. "The problem is that industry's investment in R & D is decreasing instead of increasing," he explains. "And that is a bad sign. It means that the role of government becomes even more important. I favour a substantial increase in government funding, but this can not compensate for the decline in industry. It should be a split responsibility, where both sides make sure that they address the problem in their own way."

Bloch points to the communist world as an example that complete government control is bound to fail. "Eastern Europe obviously hasn't done very well," he says. "If scientific output doesn't make its influence felt in the economy, then only the system is to blame as science is not something that goes on in an ivory tower. Scientific output is a highly important commodity, which must find its way into a country's economy and determine its competitiveness. If the country is not competitive, the system is inefficient. It was not a lack of science and engineering funding that was the problem in the USSR, but the translation of science into society."

Lessons from Japan

Bloch feels that the Japanese have been particularly good at
converting science and technology into marketable products. "The
Japanese have taken advantage of scientific developments in the
United States and in Europe. Ideas found their way to the market
place more quickly, and there is much we can learn from them. But
having caught up with us in most areas, they now have to move on
and do their own basic research. They don't appear to be working
seriously on that."

He is not in favour of putting pressure on Japan by means of
diplomatic and economic measures. "Their own self-interest has to
make them pay attention to basic scientific research," Bloch says.
"Although it's easier to be single-minded and focus on converting
science into products, sooner or later they will find out that they
have to split their resources. They must focus on both the accelera-
tion and the exploitation of knowledge because they interact."

"People would like to believe that the scientific process starts
with a bright idea, the invention, which is then finally converted
into a product. But feedback is equally important. Take for instance a situation where a certain product is desperately needed. If
there are problems with that product which you are trying to solve,
then this process can influence basic science. Scientific development
does not follow a linear mode, and this dual process has been
recognized for a long time here in the United States and in Europe.
I think that the Japanese will suddenly have to make fundamental
decisions. Their resources will never be sufficient to perform both
aspects adequately, so they will have to make compromises. And
then the world will be a different place for them."

Bloch is familiar with the startlingly blunt and sobering book,
'The Japan That Can Say No', by Akio Morita, Chairman of Sony
Corporation, and Shintaro Ishihara, a high-ranking Japanese
government official. The authors state that the United States should
accept its status as a second-rate economic and manufacturing
power, while recognizing Japan's dominant role. The publication
has therefore stirred interest and controversy in both countries.
According to the authors, the United States have become too
dependent on Japanese technology and do not have sufficient

economic, technological, or moral power to criticize Japan's industrial superiority.

And if it continues to criticize, the authors argue, Japan should take its business elsewhere and perhaps even upset the balance of power. They describe how Japan could cut off the supply of computer chips to the United States, which are vital to many advanced weapon systems. "That could be a realistic threat and we should take them seriously," says Bloch. "But we should first take ourselves seriously and ensure that our strategic industries keep up-to-date. I find it inconceivable that the United States are dependent on Japan and that we don't have our own computer chip industry. We ought to be a leader in this field as the chip was invented here. Loss of the industry is not only an economic but a strategic issue, and we can not let it happen. The book makes the right point, although we may not like the message. It highlights a real danger, concerning not only the United States but the whole world. Upsetting the balance of power is a threat to world peace."

Bloch finds it interesting that everybody listens as soon as two Japanese citizens criticize the United States. He and many others have been stressing the same point for years. "It's not too late to change, though," says Bloch. "Japan started with nothing. They had no chip or computer industry, but reached their current position through hard work and systematic education. The United States can still regain what they have been losing. We just have to make the right decisions for defence, strategic and economic reasons, and the sooner the better. We should also bear in mind that the economy's health is a strategic consideration too, and dependence on any country for our geo-political security must be avoided. If we can't retain key industries with our own resources, we shouldn't expect anybody else to retain them for us."

One of Japan's main advantages, according to Bloch, is not having had to invest in military research. "We have undoubtedly been investing for military as well as for civilian and economic reasons. Japan consequently had an easier job. They have also focused more on certain technologies and disciplines, whereas we have been blanketing all of them. Japan doesn't spend that much on health research, or on environmental research but proportionally more on physics, engineering and chemistry. This has caused some problems."

"But the focus of research is changing in the United States, and perhaps the reforms in Eastern Europe mean that things will change even more. The proportion of the US Federal R & D budget allocated to defence has decreased from a high point of about 69 percent to 61 percent today. I think it will have to decrease even further to make the balance more equal. But we are better off than we were three or four years ago. Civilian research also finds its way into defence applications much more than it did some forty years ago, when it was the other way around. For that reason alone, we have to focus more on civilian research."

Sources of tension

Bloch continues this avenue of thought by touching on the subject of military tension. In general, he favours confidence-building measures and international cooperation, and is convinced that science can play an important part, for instance in verification projects. "I think the type of situation where a private terrorist organization places a weapon in space in an attempt to blackmail the world is rather improbable. What I find less improbable is that smaller countries will have nuclear weapons which they may use recklessly. I think that is a real threat and the international community has to do something about it. Just what can be done, I don't know. That is something for politicians and diplomats to solve."

"International pressure should be able to diffuse some of the tension. And of course we should be very cautious about exporting technologies to countries which we consider may act recklessly. In reality, it's almost impossible to stop them acquiring the necessary technological details as we are in the information era. The other side of it is that we will never know to what extent the opening of Eastern Europe and the changes in the Soviet Union have been technology driven. I tend to think that access to information networks, television, computers and fax machines uncontrolled by the state, had a tremendous impact. Unfortunately, we'll never know the whole story. As I said before, there are two sides to every coin – science can contribute to human well-being, but can also be very harmful."

The NSF Director plays down the importance of another form of tension, that between demanded rights of access and proprietary rights to information. "This kind of tension has always existed," he says. "I don't think the situation has changed very much, although it might be more obvious now because of increasing dependence on technology and science. But we'll always face it, because new inventions will always have proprietary and general interest aspects. Researchers need information about what is going on elsewhere in the scientific field. At the same time, industries trying to acquire patents will put restrictions on the results of scientific research. Tension then of course develops between government and industry. Governments initiating certain measures, such as protecting the environment, might find that industry is unwilling to provide the necessary scientific details for commercial reasons. This kind of problem can't be solved and we shouldn't try to. Life is not simple, and there is after all something to be said for tension. It energizes the system and brings out some of the best on both sides. A tension-free society is a sure way of killing new ideas."

Freedom of ideas

Bloch is unable to prevent himself from smiling when describing one particular project, a search for extra-terrestrial intelligence known as SETI. After decades of dreaming, and sometimes scheming, a small band of American scientists received Federal money to embark on a US $100 million, 10-year project aimed at finding aliens in space. The sky will be systematically scanned with dish-shaped antennas to listen for faint signals from advanced civilizations which may dot the galaxy. "It certainly wouldn't be my type of project," says Bloch, "but not all science is created equal."

"Some projects are more useful than others, and I doubt that the hunt for aliens in space will be very useful. But then, Michael Faraday worked on what turned out to be electric fields. Nobody had any idea that this would result in light bulbs, in electric power, in a whole different world. He didn't foresee that, but that's where it led to. What I'm trying to say is that even if nobody sees any potential in a certain project, it still deserves a chance. We should always encourage the scientific mind so that ideas can flow freely."

Janny Groen

Research and development, stresses Bloch, should be a split responsibility, where both government and industry must make sure that they address the issue in their own way. Dr. Harry Beckers, top scientist of Royal Dutch / Shell, one of the world's largest companies, also thinks it is important to keep the differences in mind. "It is a sort of axiom from which answers to many questions can be deducted," he says. "When talking about cooperation between universities and industry, for example, it should be realized that their objectives are completely different."

HARRY BECKERS, the industrial scientist

What price research?

As Research Coordinator of the oil company, Dr. Harry Beckers looks at science primarily from an industrial point of view. Outside his company Beckers has been involved in science policy from a different angle as well. As a president of international organizations like EIRMA for example, the European Industrial Research Management Association, or IRDAC, the Industrial Research and Development Advisory Committee of the European Community.

An enormous confusion exists, according to Beckers, about the difference between science in an industrial context and science in an academic context. "On the one hand there is basic science. It is mostly sponsored by governments and done at universities with education as its prime objective. Since this can not be done without being in the forefront of science, universities have to do fundamental research. This knowledge is publicly owned and I hope it stays that way. Then there is research and development and technology produced by industry. If you look at the whole structure of the West, we are only doing research in industry in order to be able to 'kill' our competitors, or at least to be better than they are."

Basic science creates a worldwide reservoir of knowledge, where every-one can go to find out what others have done. Just as in the industrial world, there is enormous competition. The ideal of the academic scientist is to publish his work and to be known as the first. As a researcher in industry, I do not want to publish in the first place. I don't have to be in the forefront of science. My main aim is to develop something that will give me the competitive edge. From competition, society gets the quality of goods they want, and so we in industry are able to satisfy the market demand, that is, the customer. As long as I can create something that will allow the company to make a better product, I couldn't care less whether it is in the forefront of science or not."

To be competitive and profitable, Royal Dutch/Shell spends hundreds of millions of pounds on research every year and employs some 7,000 people for this purpose. In 1988, the Anglo-Dutch oil company spent a total of 428 million pound Sterling on research and development from net proceeds of 44,003 million pound Sterling. Ten years before, the R & D costs were 178 million pound Sterling from net proceeds of 19,810 million pound Sterling.

Beckers, who was born in 1931 in Maastricht, in the south of The Netherlands, joined Shell in 1955 as a research physicist. He worked in physical technology, rheology and improved methods for the production of petroleum. He graduated in technical physics from the Delft University of Technology, and obtained his doctorate from the same university. At the Royal Dutch/Shell Laboratory in Amsterdam (KSLA), Beckers was Head of the Physics Division and later of the Physics and Mathematics Division. In 1970 he became Head of Shell Planning and Strategic Studies Division in London, and subsequently of the Organization Services Division. In 1977 he became Group Research Coordinator.

Governments and industry

Governments and industry have different objectives, Beckers says, and therefore should play different roles. Governments can and should have a science policy directed at education and fundamental research. But they should not interfere in research in industrial sectors. "There are enormous dangers if politicians feel they have a role to play in that aspect," he stresses.

"Governments have a duty to promote fundamental research, done worldwide to build up the reservoir of knowledge. And you will see that there is a balance in the amount of money spent on research by the various countries. A number of years ago, for example, the West felt that Japan was not putting enough money into fundamental science. Recognizing this, the Japanese government is now making more effort in this direction."

"It often takes a long time, but in the end research spending is always in balance. If a country doesn't pay enough for fundamental research, the academics and the research establishment in that country start showing statistics from other countries in an endeavour to convince the politicians to increase research budgets. They say: 'Hey, we are not spending enough.' The reverse is also the case. If a country is spending more than other countries, the politicians will use the statistics to inform the academics that expenditure should be reduced."

"From my point of view, the task of governments is to make sure that fundamental research is done, and that education is provided for the people industry needs. Demographic changes will hit some Western countries very hard in the next five to ten years as they will not have a large enough pool of skilled people. Gradually governments and industry are becoming aware of this. They should do something about it now."

"Sometimes governments feel they should assist industries and industrial sectors in competition. They think that they themselves have to go into companies. You can of course have different views about this, but I am inclined to say that this is better left to industry. If an industry is successful, then the companies involved will have the money to do research whenever they see a potential opportunity for the future. When they don't see an opportunity for the future, they will not undertake research."

"Governments are the last to know where the opportunities are. There is a very simple human factor involved here. When your own money is involved, you really want to make sure that the project can be successful." Beckers does not blame just governments for interfering with industry. "In certain sectors, industry itself asks for intervention," he points out.

"Over the years I have become more and more convinced of the following theory: because you are competing, because you are doing research to tackle your competitors i.e. to be better than them, you would like to do twice as much research as they do. But research has to be financed. And, as a consequence, your competitors then spend half as much as you do. So you would have a competitive disadvantage, your profit is not good enough, your products are too expensive and you are losing out. On the other hand, if you were doing half as much as all of your competitors, you would very quickly see your competitive edge go down the drain."

"So you find in the various industrial sectors some sort of adherence to a certain level of research considered alright for that sector. We are all closely watching each other on that."

Upward spiral; downward spiral

"Something like inflation and deflation could happen," Beckers
goes on. "If a company suddenly spends more than its competitors
on research and development, then one of two things can happen.
The other companies don't follow and then the spender is dragged
back, forced to spend less. Alternatively, the other companies increase their expenditure on research, killing each other with competition. Their products become more and more innovative as the
companies go on spending. I call this the *upward spiral*. In the end
they are all caught in the spiral. They are spending too much on
research and their return on capital falls. Then industries start
looking around for help from government, which is more or less
forced to take out its wallet and assist. This situation has developed
into a ridiculous stage in the electronics industry. As a consumer, I
feel frustrated when a product I have purchased is obsolete within
three months because of all the innovations."

"Then there is the *downward spiral*. An industrial sector that
produces more and more of the commodity type of product, doesn't
want to do much research. Spending on R & D by the various
companies in the sector goes down and down and down. In the end
the whole industry becomes obsolete. As the companies have been
fighting one another fiercely as competitors, they don't have the
money to get themselves out of the situation. And again they go to
the government."

"We have seen this in the steel industry in the past. Governments
gave financial assistance, but I wonder whether their 'help' really
helped. I think it is far better to let nature do its job. Governments
tried to set up cooperative systems in order to provide for R & D.
But as I understand, several entrepreneurs in the United States
came up with new processes. The Japanese were involved too. By
producing higher quality materials, they were able to pick the
market up, and gradually the others were forced upwards too. Now
the trend is for R & D spending to increase again in this industry."

"Look at the computer industry. Entrepreneurs are creating new
companies but not doing R & D. They are just making clones of the
machines produced by the companies that are spending a lot on
R & D. This is putting a lot of pressure on R & D spending in the
whole of industry."

Beckers stresses that he just uses the steel and electronics industries as examples because "they are well known in the public eye – I have nothing against them." He also produces an example from his own sector.

"Years ago in the polymer-industry, people said you were crazy if you thought about a new polymer. The industry was happily producing the well-known commodity polymers, such as PVC. Now with the environment issue and market segmentation, materials research is more directed to specific uses, and there is a growing tendency to produce new materials, including new polymers. More research is being done by individual companies, who spend more on R & D for higher value-added products, fine chemicals, and the like."

Government in industrial R & D

"These dynamic movements are happening all the time. There are certain dangers if politicians are involved," says Beckers with concern. "Firstly, for those at the bottom of the research spiral, government aid won't help. This movement is a natural process which a government can not steer unless it wants to shoulder the whole global load of the particular industrial sector. Secondly, cooperative R & D programmes need enormous political involvement, so you get by definition an enormous bureaucracy. The people who determine the type of projects to be initiated are not putting up their own money. I saw this when I was involved in the nuclear industry in the United States. Whenever the Department of Energy came forward with what appeared to be quite a creditable proposal, the politicians were the first to examine it. The Senator of state X could not afford to get less money for his laboratory, and the Senator of state Y had already been given money for his reactor for so many years and couldn't be left out either. When the politicians were finished there wasn't much left of the proposal, other than just dog's food."

"Sometimes it's even worse. Governments neglect sometimes their prime task of education and fundamental science to keep the country's skills going while they do interfere in industrial R & D. The situation could arise where there is a reversal of roles with government becoming more responsible for industrial R & D, and industry being asked to participate in and provide for the money for education. I think that is putting the whole thing upside down."

Beckers likes to explain himself by drawing sharp distinctions and making absolute statements. But as soon as he has sketched the outline of his ideas, he starts adding the shades. Governments should not leave scientific research and development on environmental matters to industry, he maintains. "This is really a task for government."

"A lot of R & D is needed to determine credible and realistic guidelines and norms. Industries for their part have to find out whether their products meet the environmental requirements, thus they have also to do a considerable amount of R & D in this area."

Industry in fundamental research

On the other hand, fundamental research is not the exclusive domain of universities and other government supported institutions, Beckers asserts. "We are doing fundamental research in our company; about ten percent of our total R & D could be said to be research of that nature. That is almost the general rule in industry. Why? University professors may be doing fundamental work to educate good people, in so doing they also produce valuable stuff for industry. We like to make sure we can transfer that fundamental work into industry. In order to be able to talk to the academic world, I have to do something myself. I can't walk in and say: 'Hello! Do you have something for me?' if I don't know what the guy is doing and even don't understand him. So we do some work to be able to be involved in the latest developments. One shouldn't be so foolish to say that it is only the short term R & D which gives a return on investment. But it is always very difficult to demonstrate to shareholders that in the longer term you get a return on fundamental research."

The idea of a worldwide reservoir of scientific knowledge is dear to Beckers, although he recognizes that in reality political barriers can obstruct the free flow of information. "Sometimes politicians try to put up barriers between scientists working in fundamental research. That is a very bad thing. I am referring to the research on superconductors and low temperature fusion. There were tendencies among politicians on both sides of the Atlantic to say: 'Hey! Wait a minute! We should make sure that the Europeans keep this for Europe, the Americans for America, or the Japanese for Japan!' People who think like that forget that whatever the amount of R & D they do themselves in fundamental science, eighty percent or more will always be done in other countries. So when barriers are set up, everyone is going to lose."

"If one talks about setting up research relations with Japan, some people object, saying: they are our biggest competitors, we shouldn't do that. I think we should collaborate on fundamental research to keep the pool of worldwide knowledge growing. But cooperation in industrial research is completely different and should not have anything to do with being Japanese or American or European. The policy in this field should be based on a company-by-company approach."

Juurd Eijsvoogel

Beckers carries the red flag when he talks about politicians feeling they have a role to play in directing research in industrial sectors. He sees enormous dangers when politicians interfere in industrial research. Bloch fears that government control will hamper innovation and will direct science into one particular sector, whereas it should be broader. Viscount Etienne Davignon, a former Vice President of the European Commission and a captain of industry with long experience of diplomacy and European politics, however, points to the need for active government participation to develop research activities, as well as to carry out industrial research. "Government definitely has a role in that respect," he states.

ETIENNE DAVIGNON, the industrialist

Links without chains

"When a number of conditions are met, politics has a responsibility," says Davignon. "Firstly, there should be a need for the research. This need should be recognized by the various parties involved, not only by the politicians, but also by the scientific community as well as society. Today everybody recognizes, for example, the need for research programmes involved with environmental problems. These problems are no longer regarded as something which only concerns ecologists. Secondly, the objectives should be set sufficiently far ahead to provide an incentive for research, but not so distant that they become abstract. If these conditions are met, additional research opportunities can be added to those which already exist. This doesn't mean research programmes should be initiated which make all others redundant."

Since 1988 Davignon has been Chairman of the Société Générale de Belgique (SGB) in Brussels, also known as La Générale or La Vieille Dame, a huge commercial and industrial holding company dominating the Belgian economy.

Davignon was born in Budapest in 1932, the son of a Belgian diplomat. He studied law, took his doctors degree in Louvain, and then followed family tradition to become a diplomat. He served abroad for several years before returning to Brussels, where he became Head of the Cabinet of Foreign Secretary Paul-Henri Spaak, and later General Director of Political Affairs. In 1974, he was appointed as the first Chairman of the International Energy Agency in Paris, an organization founded by Western oil consuming countries in reaction to the sudden rise in oil prices. In 1977, Davignon began the first of two four-year terms as a Commissioner of the European Commission, at first controlling the portfolio of Industry and the Internal Market, and later adding to that Energy and Research Policies. Davignon was the architect of the European Community's plan for the restructuring of the steel industry. He has also played an important role in pushing European technology programmes like ESPRIT, the information technology project. Davignon currently presides over the Société Générale conglomerate in its Belgian palatial eighteenth century headquarters in the rue Royale in Brussels, next to the Royal Palace. The organization has stakes in approximately 1,200 companies, and its activities range from banking, cement manufacture and diamond trading, to transport, chemicals and the production of golf clubs.

"It is never enough to simply have a good idea, or even to just prove that it is true," stresses the industrialist. "It has to work. The whole concept of bringing together universities, businesses, information technology and various administrative bodies has obviously worked, otherwise there wouldn't be such a large disproportion between the number of projects proposed, and those that can be funded. This basic change in infrastructure has meant that 200,000 people have learned to work together, who otherwise would not have done so."

When European Commissioner Davignon set up a 'think tank' to assist him in selecting targets for future development. "The responsibility of this completely independent group of major scientists was to tell me, or the institution, what they considered needed to be done to anticipate changes in science and society." Davignon is currently trying to form a similar advisory group of scientists for the Société Générale. "I believe very strongly that we must bring our scientists together, even if they are active in different fields. They all belong to the same organization, and they can benefit from one another by thinking along the same lines."

Forging links

The new role which science is playing in the Société Générale has much to do with the major changes which have been taking place at the rue Royale since 1988. In that year the Italian industrialist Carlo de Benedetti launched a hostile bid for the huge business empire. The dramatic take-over battle kept the European business establishment in tension for six months, and ended when the French investment bank, Compagnie Financière de Suez, was called in by SGB as a White Knight and won a controlling stake. When peace had been finally signed between the large, competing shareholders of the Générale, Davignon was appointed Chairman. A new management was then established to mould the widely differing elements of the holding company into a cohesive group.

"By bringing our researchers together, I think we can improve the technologies which exist within our company," says Davignon. "We can also determine how to achieve a competitive or comparative advantage that will enable us to expand into other areas, which is one of our aims. It's important to organize meetings between different researchers within the group, and to ask all sorts of questions. It's equally important to let your staff know that these discussions are an inherent part of their task and not some exotic activity. If they think that the upper hierarchy is not interested, they will not come up with anything."

Davignon also likes to exchange views with people in comparable positions in other industries. The European Round Table of Industrialists, a prestigious but informal group of leading European industrialists, provides an excellent forum for such discussions. "Training is now a topic that comes up nearly every time. How do you get the people with a sufficiently broad approach. There is at present a very high degree of specialization in many disciplines, but specialization is not the answer to our problems. It is indispensable, but it is also insufficient. Those coming directly from schools and universities are not adjusted to the world in which they have to live. The world is everyone's world, and not your world."

"For a man in my position," Davignon says, "it is vital to be well informed, and particularly to keep in touch with major scientific developments. Finding adequate information, however, even in the information age, is a considerable problem. It is especially difficult for a non-scientist to obtain an insight into trends, major breakthroughs and possible applications."

Davignon finds that the mass media do not provide him with what he needs, and neither does the specialized scientific press. "Not because they are not doing their job," he explains, "but because general information is no longer sufficient and specialized information is not usable. There has to be a balance between the two. Specialist information is not digestible to people in my position because we are not clever enough or do not have sufficient time. It really comes down to a question of self-organization. I am personally very fortunate, because I have met a lot of people and have established a number of contacts so that I know where to go when I want to find out something. The appropriate information never comes to you, you have to find it."

"This is also true in another way. Nowadays, there are consultants for every conceivable subject. But the value of a consultant is only as good as what you have decided you want to know, and what you then decide to do with the information brought to you. You have to select what is relevant yourself. If you are unable to carry out such a selection, you can not utilize what is available. Adequate information is the target, but we still have a long way to go."

"The new information technology is transforming every aspect of life. A computer is regarded foremost as a better tool. But its users do not grasp the qualitative change. A word processor tends to be seen as an important form of typewriter, which it is. But not exclusively. So long as we fail to understand all the additional functions or how they can help us do things differently, we will continue to use only 10 percent or 15 percent of the additional capability available. Nobody seems to be making use of the information technology to its full capacity. The great problem we now face is how the technological explosion should fit into our way of living. Science and technology are not solely for the use of specialists in certain areas, but should become features of daily life."

Stimulating research

In his position as a European Commissioner, Davignon planned the restructuring of the steel industry and played an important part in encouraging European technology programmes. Both as Commissioner and as a member of the European Round Table of Industrialists, Davignon has repeatedly stressed the importance of developing a European high-tech industry by stimulating international research programmes.

"I was very struck by the difference that existed between Europe on the one hand and the United States and Japan on the other. In

America, for example, there are no barriers between science, the business world and the public world. Such barriers unfortunately do exist in our countries, and I wanted to do something about it."

"As far as fundamental science was concerned, the quality of European science was comparable to that elsewhere. But there was a danger that the ability to maintain this quality would be affected by the way in which universities tended to be isolated from other sectors of society. This would undoubtedly create problems, because the importance of science should be that it can help anticipating future situations."

"It seemed absurd to me not to include cooperation in the fields of science and technology in a European integration project. These dimensions are vital if integration is to succeed. Science is clearly international and scientists don't tend to feel that research findings should be solely kept for use in their own country. Scientists don't have the national hang-ups that you find in other areas."

Davignon does not feel that the intention of stimulating European research and providing financial support, is to combat scientists in other trade blocks or to retain discoveries exclusively within Europe. "The comparative changes of a scientist in Japan, Europe and the United States must not be different, otherwise a brain drain will be created from one region to another. That could lead to research concentrating in a small number of places, which is not favourable to science. In certain fields of science, for example electronics, Japan has clearly gained such an advantage that it is pushing Europe and the United States out of that field. This might create a knowledge or technology gap, which must be avoided."

"But science never stops. It should be accepted that, for instance, the Japanese are currently stronger in the field of semi-conductors than any other country. But that doesn't mean that the next generation of this technology will be Japanese too. That's why there is often collaboration on big projects, such as Philips and Siemens working together on the Megachip Project. Catching up is very expensive. It takes leadership to convince people to pay that high a price."

Juurd Eijsvoogel

Davignon stresses that science is clearly international and he is vehemently against scientific protectionism. According to the industrialist, scientists in general do not tend to feel that research findings should be solely for use in their own country. Nobody would agree with him more than Robert Solow, a staunch advocate of free enterprise, who won the 1987 Nobel Prize for economics. Solow says that he can not emphasize often enough the importance of science being completely international and accessible, so that everyone and every country, be they poor or rich, should be able to tap into science sources. "I suppose that you must protect military secrets, although there is a danger that the military people begin to think that everything is secret, and that can't be allowed to happen. It's the work of our politicians to control the military."

ROBERT SOLOW, the economist

International openness

No protectionism. Ideas and goods must flow freely, is Solow's adage. Still he was mentioned by a leading American business magazine as being one of those diehard free traders who are changing their tune. Under the headline 'Swan Song for Laissez-Faire', it pointed out Solow as one of the free traders having second thoughts about the validity of that economic perception. According to the magazine the eye-opener for Solow came with the realization that the electronics age has yanked the rug out from under a basic tenet of economic theory. Information technology and global tele-communication have led Solow to conclude that the theory of comparative advantage is obsolete.

In his rather untidy little room, stuffed with books, in the Massachusetts Institute of Technology (MIT) in Boston, Solow agrees that science and technology play a dominant role in modern economics. Vehemently denying having second thoughts about the theory of free trade, he declares he always has been and still is a diehard free trader. Born in Brooklyn, New York, in 1924, Solow graduated from Harvard University in 1949. His special fields of interest are mathematical economic theory of capital and growth, macro-economics and the economics of natural resources.

Elected to membership of the US National Academy of Sciences in 1972, Solow is the author of several books on economic theory. He served on President Lyndon Johnson's Commission on Technology and Economic Progress and on John Kennedy's Council of Economic Advisors. As a Nobel Laureate he was appointed to Vice-Chairman of MIT's Commission on Industrial Productivity to assess the United States' industrial competitiveness. In May 1989, after two years of study, the Commission released 'Made in America', which catalogues the problems distressing the American economy and which made headlines because of its rather critical tone.

Changing concept of comparative advantage

"I think that magazine article missed my point," Solow explains. "When economists ask why do some countries produce certain goods; why do the Dutch produce cheese and electronics and the United States automobiles and washing machines, the answer has always been in terms of the theory of comparative advantage. When I was a student, the assumption was that a nation's comparative advantage depended mainly on its natural resources, climate, location, nearness to raw materials and markets. This probably was true up until the earlier part of this century."

"Today, much trade, at least between developed, advanced countries, is in highly technical, manufactured commodities, and in very skilled services like banking and insurance. The advantage that one place has over another to produce a particular commodity depends very little on natural resources or on location. Comparative advantage is created by a nation or an industry of that country. To have the comparative advantage in the production of a particular type of electronics, for example, is to be first in the market, to learn to reduce costs from the experience of production, to enlarge the market so that an adequate amount can be produced, and to defend that position against rivals by innovating rapidly and improving the product design. This doesn't mean that the notion of comparative advantage is now obsolete, it simply means that the way of acquiring the comparative advantage has changed drastically."

"It is no longer enough to say, for example, we have iron ore, therefore we can produce steel. Having the comparative advantage is being able to say: we have mastered this industry, we were first in the market, we dominate the market, and we will do what is necessary by way of new products, methods and design to maintain this position. Since being the best depends heavily on the skills of workers, designers and engineers, the comparative advantage can be achieved by education and training, by cultivating human abilities."

Solow says he knows that a new economic school of thought argues that some of the underlying assumptions about comparative advantages have been wiped out, because of Japan's 'targeted industry-strategy' for example, or the ease with which technological know-how and capital can be shifted around the world. The economic point was that an American company can spend millions of dollars developing a new product, only to watch helplessly as imitations flood the market before the product has reached break-even. Copycat producers can always undercut prices because enormous development costs do not have to be recovered. Unlike these economists, Solow does not believe that these assumptions have changed economic theory.

"These assumptions have, however, caused a change in what economists and business people think about. Just as the discovery of new classes of materials doesn't change chemistry but the way chemists and physicists think. One of the problems of the American economy for the past fifteen years has been an unwillingness to imitate foreigners. Americans feel that they have nothing to learn from manufacturers, engineers and designers in other countries. Part of the process of creating a comparative advantage in the production of high technology, is the ability of companies in one country to learn almost instantly from those in other countries."

The professor looks up pensively as he says that Americans have a special problem. "They do not speak other languages. American industry has only just began to learn that it needs people who are at home in other countries and cultures. If they have to sell to other countries, then they must know what the people are like and want to buy, and learn something about the laws and social institutions of these countries. Teaching of foreign languages and cultures has been diminishing instead of increasing in the United States, but people in industry and government are only now becoming aware of this problem."

Cooperation instead of protectionism

Solow strongly opposes any kind of protectionism. Some Americans argue that, because they are good at invention and because modern technology makes it easy to copycat, bright ideas should be protected. This is not Solow's approach. "The proposition is true that the United States are still very good at basic science and fundamental inventions," he affirms. "We are not so successful, however, in converting a fundamental invention into a commercial product, which people want and which can be produced and sold. Protecting inventions is a sure way to dry up the flow of new ideas. We in the United States have to commercialize our inventions as well as those of others. At the same time, we should urge other countries, especially Japan in this case, to devote more resources to basic research."

"In 1987, when I won the Nobel Prize, one of my colleagues here at MIT won the Nobel Prize for medicine; Susumu Tonegawa from Japan. Shortly afterwards we had a lunch here in Boston, which was attended by visiting executives from a large Japanese company. After lunch, the Chairman of the Japanese delegation made a speech in which he said it was wonderful that Japan and the United States have reached this very comfortable arrangement. In the United States we did basic science and in Japan they turned it into a saleable product. When he had finished, the master of ceremonies asked Tonegawa and me to speak. Tonegawa said that as a scientist he had been driven out of Japan, because he could not get the resources to do fundamental research. He thought it essential for Japan to do basic science. I in my turn said that the days were over when the United States could not pay attention to commercializing discoveries and that we had better get on with the job."

"This is the right way to react. Instead of taking refuge in protectionism, the two countries must cooperate. Japan can no longer say to themselves, we will live off American and others' inventions. And we in America must stop saying we are the greatest people in the world." Solow is glad to observe that the attitude of both Japan and the United States is changing slowly. "That's healthy," he remarks.

Consequences of a military budget

The Nobel Laureate points out that Japan has had a great advantage over the United States in not having had much of a military budget. "It is interesting, but not an argument," he says, "that the two industrially most successful countries of the post-war period, West Germany and Japan, have had the smallest military expenditure. Whereas the United States, Great Britain and France, which have been less successful, have all had much larger military budgets. It is not the budget by itself, but the large proportion of skilled people, the very best scientists and engineers, which has made the difference in these countries, especially in the United States. There are cases of the results of military research spilling over to the civilian economy. The jet aeroplane is the single, most important example. Much of the IBM computer revolution was for military purposes and has since revolutionized civilian industry."

"These days it appears to be the other way around. The most advanced technology in civilian industry is now spilling over into the military. But to do this, the military establishment and defence industries absorb much excellent scientific and engineering talent. The cost is hard to measure but it is a drag on the American economy which needs to be changed."

Some people at MIT are organizing to try to convince people that a conscious effort should be made to devote some of the military research budget to other types of projects, to environmental research for instance. Some of the budget should also be devoted to commercial and business research. "This will happen. The only question is how quickly; the sooner, the better," proposes Solow.

"Can the United States afford not to make these changes? Yes. It is a very rich country and can afford to wait, but it will be costly. It is better to do it right away."

Management and technology

Solow shakes his head in disagreement with some observers who say that the United States does not produce enough scientists. Pointing out that Japan turns out scientists at twice the per capita rate, they say that the United States Government should encourage young people to study science and technology instead of business. "You simply can not decide this from the top," he asserts. "If there is a demand for business administration degrees, then people will study for them. What you can say is that business education in the United States has adopted an incorrect belief that business management is a skill that can be taught independently of the type of business to be

managed. It is believed that a manager can manage a steel mill, a
ship yard, a coal mine, a bank, or anything. I think this is a
mistake. If I am right, we need many more scientists and fewer
managers. We need people who have learned something about the
technology of a particular industry and about management."

On the subject of federal assistance for scientific industrial projects, Solow reacts with caution. A lobby group founded to promote
an industry led policy, 'Rebuild America', recently proposed close
government and business collaboration with industry deciding which
technologies should receive federal assistance. "Allocation of
government assistance of scientific and engineering research is a
very complicated matter. The common belief in the United States is
that the government will not be good at predicting which industries
will be successful."

"Why not let scientists decide which projects deserve federal
assistance? There is a tendency for the choice of specific projects to
be motivated by technology. This is a particularly beautiful invention, therefore we should invest money, that's how scientists react.
But from the point of view of the national economy, that is not the
right way. The beauty of an invention has little to do with its
usefulness. A good indicator that a particular line of research is
likely to be of great economic significance, is that the businesses are
willing to spend quite a lot of their own money. If this is the case,
then often it's a good idea to help them along."

"So the American solution will probably be that the government
provides some funds, subsidies and resources and industries have to
compete for them by showing a willingness to invest their own funds
along with government money, or in some other way. There will
have to be some kind of open competition."

Research consortia

One way for American companies to compete with foreign firms,
which are extensively funded by their governments, is to create
research consortia, says Solow. The anti-trust laws exclude such
consortia from the United States. Solow argues that they should be
allowed, "because this will enable smaller companies to undertake

modern research, which is often too expensive for a small electronics or software firm on its own. Research consortia may even in some ways increase rather than reduce the amount of competition."

Solow is less happy about international consortia, not simply because he thinks national boundaries ought to be drawn. "You must remember that the American ideology is that competition is a good thing. One of the ways in which I can justify to myself allowing consortia of American companies, is that there will always be foreign competition. The danger is that firms which cooperate on research will not compete with the final product but rather monopolize it. If a group of American companies gets together and discovers a wonderful new product, that danger is limited; they can not really exploit the consumer because sooner or later there will be competition from foreign imitations."

"If we invite also foreign companies into these consortia, where will the competition come from? I have no objection to international consortia as long as there is a source of competition. The trouble, however, is that a consortium of companies from different countries will divide the market up amongst themselves. For example, the Dutch company will have the Northern European market; the French company France, Italy and Spain; the American company North America, and so on. That's not a healthy situation. The American view is that competition protects the consumer against the producer."

Solow does not agree with the notion that trying to block the formation of international research consortia might encourage science and technology wars. "I think that's the wrong way to put it," he explains. "The trouble with a military war is not that it is competition but it is destruction. I do not object to the Olympic marathon, although you could regard it as a running war. Let there be more active competition in science, engineering and industry, as long as it is not destructive. This is unlikely because science is one of the least national activities. Most science is quite international. Half of the letters I write about my own profession are addressed to people not in the United States, and half the letters I receive are not from Americans. The community of economists, chemists, physicists and electrical engineers is truly international. So I don't fear that kind of competition at all, and least of all I would think of it as war."

Closing the gap between rich and poor nations

"I fear the growing gap between developed and underdeveloped countries. It will be hard to bridge because poor countries do not have access to science and technology and can't afford research and development. This is one of the most difficult policy problems. The most obvious solution is whenever there is the merest beginnings of an industry in an underdeveloped country, it should be protected and given a chance to grow and establish. Only when it is in good shape, competition from abroad should be allowed. In principle, there is nothing wrong with this, but in practice, it is very difficult ever to come to the stage where you say we are now ready to face international competition."

"So I think the first and most important thing that the rich countries can do for the poor countries in this respect, is to give them complete access to their markets. Make it very easy for a small firm in Mexico or Brazil to sell in the United States, the Federal Republic of Germany or The Netherlands. Secondly, as we already do, make it even easier for poor countries to send their bright young people to the United States or Europe to study and to learn about science and engineering. There is of course the problem of the brain drain, for which I have no easy solution, I must confess, apart from appealing to young people, that having been educated at the expense of their native countries, they owe some years of their lives. However, for them it is very attractive to stay in the West. The brain drain is indeed a very serious problem."

Solow does not think it is a good idea to send scientists and technicians as development workers to Third World countries. "It is not the isolated teacher that makes a difference," he says. "A university such as MIT which has a thousand teachers could not be broken up into one thousand one-person universities. There is something about the group and any important subject requires a great variety of experts, so they all need to be in the one place. The time may come, when West Africa has a good engineering school. Not in Nigeria or any single country, but perhaps a federation of states working together in science could create a technical university. We should encourage and help them by training their people and sending some of our own people to teach in these universities. The setting up and organizing of such universities is bound to be a public function in which the United Nations could play an important role."

As for the question where the openness of basic sciences does end and the specific know-how, the competitive edge in the technology for production takes over, Solow answers that business firms clearly will want to keep for themselves their own technical skills. "Wherever the boundary is drawn," he argues, "there will be some close cases. Important, however, is that business secrets be genuine and not mere scientific secrets and that distinction is clear enough. Science has to be a completely open activity, otherwise it will die."

Janny Groen

The arrangement that the Japanese find so attractive, in which the United States do basic science and Japan turns it into saleable products, does not please Solow at all. In order to create a more equitable balance in the world, the Nobel Laureate, joined by the director of America's National Science Foundation, urges Japan to devote more resources to basic research. Professor Hisao Yamada, director of Research and Development of the National Center for Science Information System (NACSIS) and Professor of Information Science and Management at the University of Tokyo, explains why his fellow countrymen are struggling with basic science. According to Yamada, Japan's cultural restrictions hamper its development into a leader in basic scientific research. "Our education system breeds us to be excellent followers."

HISAO YAMADA, the scientist in Japan

Breaking the mould

"The achievement of Japan is for the most part the result of hard working, not of hard thinking," Professor Yamada says. Though Yamada admits that it is rather un-Japanese to express critical notions on his own nation, when discussing the relation between science, society and the role of communication in a Japanese perspective, he feels the urge to first explain the peculiarities of the culture he is living in.

Moreover, Yamada feels uniquely apt to do so, after eighteen successive years of scientific experience in the USA (at General Dynamics, at IBM and at the University of Pennsylvania) and – following a stay in Japan – two years at the University of Delaware and Stanford University. The last ten years he lived in Japan permanently.

"Science in Japan is different," he says. "In our society values like creativity and individuality are carefully filtered out, although not by intention. Our education system breeds us to be excellent followers, to excel in imitating others and in improving tried concepts. As a result, Japan is good in application oriented sciences, such as engineering. In basic sciences, we are not so good. The reason is that developing 'basic ideas' is not so simple in the Japanese society. It contradicts our culture."

Excellent information

Though Yamada perceives Japan as better in imitating than in initiating, Japan surpasses the rest of the world in a growing number of areas. The existence of his own organization NACSIS

(National Center for Science Information System) is but one example of Japan taking the lead. So far, NACSIS is the only example in the world of an extensive and highly sophisticated computer network for science information, which links universities as well as individual researchers directly. NACSIS participants can at low cost retrieve the latest scientific publications (mostly as abstracts but in some disciplines as full text) and can update themselves on the progress of ongoing research and similar 'grey literature'. The network allows electronic messages and informal communication between scientists. NACSIS, first developed in 1973, has now registered thousands of Japanese researchers who have direct access to scores of databases. With a nation-wide catalogue and document delivery service, it links over 130 university libraries in Japan and is still growing.

Yamada sees no conflict in the paradox that an imitator can by itself never become number one. "We learn to do things second, with the aim of outperforming the one who was first," he explains. Other countries had long before started to work on the concept of one nation-wide information network for science, he says, citing the American initiatives to link on-line campus libraries (such as OCLC). The success of the network in Japan is due to the power of the Ministry of Education, Science and Culture who gently forces the universities to participate and because of a lower level of resistance against centralization and uniform planning in Japan in general.

Yamada's face brightens momentarily with a warm smile. "Besides," he says, "excellent information supply, knowing exactly what others are doing elsewhere, is a necessary precondition for perfect imitators. They can not live without that. Picking up the right information is the best way to advance over others. The guy who best capitalizes available information, gets the cream of the cake."

Japanese success factors

Yamada points to: "Two or three basic facts that make Japan economically successful. First, we are a uniform society – and not by accident. It is the official purpose of the powerful Ministry of Education, Science and Culture to develop an intellectual labour force whose main training is to be industrious and diligent. Our children are drilled and our school system is imbued with characteristics of uniformity in this sense. The fault tolerance and allowed deviation from the standard is an extremely small margin. Creativity is punished – our son who had been to school in the USA for some years, got into serious problems with his teachers in Japan because he kept asking 'why' in class. The question 'why' is not interpreted as a healthy sign of curiosity – it is a personal offence towards the integrity of the teacher! We finally had to transfer our son to a very expensive, private school where children are allowed to ask why."

Uniformity

"The trait and requirement of uniformity is very visible at universities too. In the USA, graduate students receive a bonus for trying new concepts in new fields and doing something new: they bounce their ideas to find a problem until they are licked. When I came back to Tokyo in 1972 I soon found out that you can not ask Japanese students to do the same. Ask them to find a new avenue and they get lost. Everybody has to march in the same direction. They are expected to improve existent knowledge by adding further details."

"Similarly, it is impossible to apply 'crazy' ideas in research or to focus your efforts on unorthodox approaches. If you apply for funding, it is not advisable to base your request on the argument that you will be the first to do it and that you might or might not reach a breakthrough! The most applicable argument for Japanese research is: 'Everybody in the rest of the world is doing this and if we do not do the same, we will get behind others'."

"In the same sense you may not question the work of your scientific colleagues. First findings may not be proven incorrect later when falsified by a colleague. That is taken as a personal failure on the part of the first person, not on his abstracted theory. This often hampers proper peer review and thus the formation of new ideas and theories, because it would frustrate the system if someone poses something that has not yet been proven again and again. Because the system can not correct itself. The result is a homogeneous and industrious labour force, drilled to work hard, with almost a ban on mavericks and new ideas."

Quality control

The second factor Yamada points to as a reason for Japanese success relates to the absence of natural resources. "When Japan was poor and did not have its own resources, we bought our raw material everywhere in the world – anywhere where merchandise was cheap. As a result, our basic material never had an equal quality. For long, Japanese products had the image of breaking down easily. The turning point came around 1970, when Japan began to compensate for the low quality raw material by improving quality control in the manufacturing process. For such severe process control it proves to be handy to have an obedient, uniform workforce. The clean room concept for the production of micro-chips is a piece of cake for the average Japanese employee. In the USA, on the contrary, companies struggle to make their people obey the strict measures for keeping any dust out of these areas."

"Right now, Japan excels in most high-tech areas where precision control in the manufacturing process is required. "This," says Yamada, "has very little to do with applying creativity or coming to complete new ideas. It is a matter of diligence."

The third factor for Japan's success has to do with its religion in which the sun is worshipped. "For Japanese, the sun symbolizes having to work. When the sun is out, it is time to work. And working is good, a pleasure, a fulfilment of life. Not a necessary evil as it is for Christians," says Yamada, referring to a Japanese saying that implies that you can never 'excuse' yourself to the sun for not working.

Yamada rounds off his explanation. "So if you take together a uniform and industrious workforce, the skills of tight quality control and of adding the utmost value to scarce raw material, as well as the deeply felt benefit of working hard, you have the ground principles for Japanese success."

Brain power

However, it might be argued that in a 'sciencified' world where economic competition is no longer determined by the availability of natural resources, but much more by the so-called 'human resources' of brain power, Japan has a clear competitive edge over other world powers. Yamada shakes his head. He does not agree. "In Japan brain power is focused on mass production, uniformity and high quality only. Flexible manufacturing techniques leading to many combinations of different features for consumer products only pose an illusion to diversified individual choice. The choice is superficial, not in the basic supply of goods."

"Brain power in Japan is limited; the nation works hard, but does not think hard."

"And if economists say that the availability of natural resources no longer determines the rank in world competition, then that holds true for only the rest of this century. But ultimately the countries with natural resources determine the play. Look at OPEC countries, how silly they are not to join their forces more effectively. They could rule the world. Japan may have rice rations for one year, but our oil reserves no longer last more than two months. If, as another example, Australia, Canada and the USA form a cartel on agricultural goods, they can easily squeeze us, as our cattle completely depends on their corn."

"Of course as a scientist, I have faith in the ultimate success of human knowledge, like the possible development of factory-made beef and protein food. But even if that is feasible within the next 100 years, it does not solve the problem. Science contributes great things to society, but is not a solution to everything. Suppose we can replace all the wood we use by certain plastics, then the production of plastic would still use up too much energy."

Danger point

Does Yamada believe that science solves or adds to society's problems? He has his concerns, but remains over all an optimist: "It is my personal belief that mankind is very close to the danger point where a catastrophe may take place in the ecosystem. It will still take us ages to restore the rain forest, even if we start today. Christianity does not stop preaching to 'go forth and multiply', while the world severely suffers from overpopulation."

"It is my opinion that it is the duty of science to convince the world now that we have reached the point where the world can only just sustain us."

"But who has things under control? Politicians, I say, are a poor lot. The best government is an informed government. But most governments are made of politicians who have to appeal to the prejudice of uneducated people."

"Confucius said 'Man of ill education will spend time unwisely'. As to the future, I am convinced that science will continue. And if we meet the requirement that we can stabilize population, science and engineering will provide us with the tools to make the best of it. But the Japanese traditional philosophy of working hard for continuous growth will have to go. Instead we should satisfy ourselves with producing enough for everybody and not always more and more for each. It is the duty of society to educate the people to use the excess time wisely, like in intellectual activity or in art."

Yamada is convinced of the beauty of science: "To me, science has an intrinsic value. Knowing is by itself fascinating, not necessarily for improving things, but just for the value of gaining knowledge."

Scientific standards

Yamada goes on: "Science is the root of modern society and originates from all cultures where life was harsh and the margin for survival was narrow. That is why scientific reasoning fits much better in a context of Christianity, which applies much stricter rules. Christianity reflects the harsh climates it comes from. Compare Buddhism or its predecessor Hinduism, for example; a lenient and sometimes arbitrary religion that combines perfectly with the gentle climates of its countries of origin. When the first Christian settlers tried to convert the Hindu Indians in Asia, the people of India thought: 'How nice; one God! Let us add that one God to the many we already have.' But Christianity does not allow such easy standards. And the same occurs in science. Scientific truth is narrowly defined; something can be falsified or not. This is different from the oriental cultures where we seek harmony of multiple values instead of an exclusive choice between yes or no."

"The power of science in the Japanese society should not be overestimated. This is a society that does not allow freaks, nor really new and crazy ideas. So far, Japan can only be good in areas where hard work is needed and patience. Where scientists concentrate on more details. Take materials science: Japanese scientists are good in discovering new compounds for synthetics. That is mainly because they have the patience and diligence to try all possible combinations, not because of the theory that lies behind it."

Even so, Yamada believes that "Japan realises that it has to boost creativity to really gain the world lead. But its culture makes it very hard to do so."

He concludes: "I believe that Japan has to change its educational system drastically if we want to produce more original ideas. The consequence will be that the average level will most likely come down. Right now, the country has an extremely high level of educated people, but very little deviation in end terms. In the USA the average education level is lower, but allows much more deviation so that the country does succeed in having some excellent innovators."

"Creativity is the endless pursuit of a seemingly unsolvable problem. If Japan wants to become good in that, we will have to lessen our cultural restrictions for our next generation."

Eefke Smit

Culture plays an important part in the way scientists of a particular country approach science, says Hisao Yamada. So does ideology, adds Professor Tudor Oltean, a former Romanian refugee born in Transsylvania in 1943 and now Professor of Communication Sciences at the University of Amsterdam, The Netherlands. He points to the practical problems of scientific life in communist countries: the lack of hard currency for Western learned journals and books, restrictions on traveling abroad to scientific conferences, endemic shortages not only of scientific equipment but even of such necessities as scribbling books, constraints placed by the home government on the ownership or use of potentially 'subversive' equipment such as typewriters and photocopiers. In David Halberstam's words: how can a country be afraid of those machines and still be competitive?

TUDOR OLTEAN, the communication analyst

East-West mergers

Scientists in Eastern Europe are not only confronted with such tangible stumbling blocks as shortages of books, journals and equipment. Oltean tells a rather ironic and illustrating story, referring to an incident which happened in Romania in February 1990, just after the revolutionary turmoil and the summary sentence of Ceausescu: the paperboys in Bucharest went on strike. To Oltean this strike is more than an incident. It marks what had happened to the information process in the ideological, centrally directed economy of Romania.

"Ideologies need an information monopoly. The communist state did everything to limit the available information. 'News' hardly existed. So when the ban on publishing 'real' news was broken, obtaining the newspapers became a mania! Demand increased dramatically and the paperboys could no longer carry that load. They went on strike."

Black market for information

As a keen observer and analyst of communication processes, Oltean (who moved to The Netherlands in 1977) describes some of the changes taking place in Eastern Europe in the information structure, both in daily life and in science. "Rigid state control leads to the devaluation of news. The goal of information is to educate people to become good socialist citizens rather than inform them of what is happening in the world."

"To still meet their information needs, people start communicating at different levels of interpretation. And individual communication channels provide a vivid 'black market' of information; for controversial political news but also for new scientific information."

For scientists in Eastern Europe, the main problems of life are practical: lack of hard currency for Western learned journals and books, restrictions on traveling abroad to scientific conferences, endemic shortages not only of scientific equipment but even of such necessities as scribbling books, constraints placed by home government on the ownership or use of potentially 'subversive' equipment such as typewriters and photocopiers.

But Oltean says: "Material that was official forbidden to be imported could be obtained on the Black Information Market. For many publications it could be a matter of time. Every next dictator starts publishing what under his predecessor was not permitted. So under Krushchev, everything could be read that was forbidden by the Stalin regime and under Breshnev we read the Krushchev forbidden list."

Banned authorship

Oltean's own work is still banned from official Romanian libraries, as a result of his refusal to return to Bucharest after an exchange visit of scientists to Amsterdam in 1977. "I no longer exist in Romania," he says.

In the centrally directed economies of Eastern Europe, science performs a different role than in the free world. Under communist regimes, Marxism-Leninism itself is a science, instrumental in building up a socialist society. The scientist himself would become a new, secular preacher for society in which other religions were declared outlawed.

But Oltean explains how scientists slip through the net of rules. For example, originally his thesis on 'The Morphology of the 18th Century European Romantic Novel' slipped past the state censorship in 1973 although it was entirely non-Marxist in its reasoning. "As an obedient Eastern European scientist, I would have been supposed to give a proletarian interpretation of the romantic period and explain it in terms of ownership, establishment and class struggle. The official version should have read that romanticism was typical for the leisure of the aristocracy. While exploiting the working class, the elite themselves had nothing else to do than falling in love!"

"However, I did not apply the Marxist interpretation. In my opinion, it is the scientist's task to be non-conformist. So I gave a non-Marxist interpretation of the romantic novel."

Right now, Oltean's scientific focus is on a typical Western phenomenon of the consumers society par excellence. He studies television soap series (Dallas, Dynasty) in order to draw the parallel with the composition of the 18th century romantic novel. In his 'study station' full of video-tapes he smiles charmingly: "Really, very little changes in romance over the centuries."

Non-Marxist science

The history of communism proves that it is hard to direct science in an ideological mould. Anti-Marxist comments can be easily controlled and picked out, but it is more difficult with non-Marxist contemplations, as Oltean explains. "Censorship in centrally directed economies has only relative power. The censors look for texts unfriendly for the state. For a non-conformist scientist on the other hand, the degree to which Marxist reasoning is absent in a text has information value; it is a recognized tool in communicating with his colleagues."

"The result for science in general in Eastern Europe is that it is far less Marxist than one would expect from its ideology."

Again he smiles: "The application of this second level of interpretation has undermined the official ideology. Go out to find the real, fully convinced communists out there after forty years and you find they just no longer exist."

From the practices in the recent past, to the future. How does Oltean as a specialist in communication sciences think the situation in Eastern Europe will change?

"It is remarkable that science in Eastern Europe succeeded in not being completely isolated from the world play. Clearly, after being cut off officially for more than forty years, there is a strong urge for getting involved in the international communication again. Mostly among scientists, as access to international communication is vital. Information has been scarce, foreign contacts were impeded, funding of research was poor, but it would be unjust to conclude that a catastrophical leeway is representative for Eastern European science. As I said before, people in Eastern Europe have learned how to slip past these obstacles."

"Some highly prestigious scientific centres exist that will catch up quickly. For historical and cultural reasons, they will look to the rest of Europe rather than to the USA or Japan, in finding a role-model for their own developments. This may shift the power relations in science worldwide towards the European continent."

And, nearly as a prediction: "Eastern Europe has its own unique characteristics to offer to the rest of the world. It will not just be a consumer zone for the next ten years."

How will more communication affect society? Oltean is firm in his conviction that "dissemination of information leads to de-ideologization of society. It undermines the difference between the rulers and the ruled. In a sense, the new haves and have nots in society could be the people who have or have not access to information, although I am prudent in picking out only one explanatory factor for the structure of society."

"But, information developments lead to social change, that is for sure. Information immediately influences the balance of power. The role 'glasnost' plays in accelerating 'perestroika' in the Soviet Union is a convincing example. Technology and the rise of networks have set in motion a quiet revolution already in the seventies. Societal change demands that more information is available to more people. Then the relationships between world enemies will change by the availability of information. The consequence is that we need a different kind of regulation, a legal framework – also internationally."

Information overload

How is science influenced by the abundance of information nowadays? "In the communication sciences, much has been said about the effect of the availability of information. Firstly, there is a natural limit to the amount of information that can be absorbed. But it is a mistake to think that there is too much information; that is an out-dated interpretation of the way information is regarded nowadays. Secondly, in a more recent wave, there has of course been a lot of attention to the disadvantages of worldwide information, especially when being in the hands of only a few mighty commercial companies. International television has caused the loss of many national and cultural traditions. In Eastern Europe people yearn for information right now, with its respective consequences too, of course. The special subculture of the 'black market for information' is very likely to disappear. And commercialization will pose its own threats to the community."

"But science knows no boundaries by tradition. In general the scientists' desire to publish goes hand in hand with the information needs of the scientific audience. More so, the developments impose new ways to make the available information of use to us."

"The paradigm of today is the Neo-encyclopedism: it is essential to have information on information, the so-called secondary information. In this technology era, databases are our modern encyclopedia. The use of primary information changes. It is not so much a matter of an abundance of information, but of the selection tools that are provided."

"If I were a scientific publisher, that is what I would focus upon."

Eefke Smit
Anita de Waard

Technology and the rise of networks set in motion a quiet revolution, says Tudor Oltean. He muses that societal change demands that more information is available to more people. All types of relations change by information. The Professor of Romanian origin argues that internationally we need a different kind of regulation, a legal framework.

Do not expect the lawyers to provide the world with such a framework, warns Rudolf Bernhardt, one of Europe's most experienced and influential judges, who has shown a lifelong concern with interpreting society's needs in the light of the law. Describing the role of lawyers, who are ultimately called upon to assess the legality of ethical issues in the light of existing legislation, he says: "Our task is to note developments, judge how the law should be interpreted, whether it is adequate and how it can be adapted to meet the needs of modern society in a changing world. But the law only follows, it only responds."

RUDOLF BERNHARDT, the lawyer

We must not lead

Since 1981 Bernhardt combines a distinguished career as Director of the Max Planck Institute of Comparative Public Law and International Law in Heidelberg, with duties representing the Federal Republic of Germany in the European Court of Human Rights in Strasbourg.

Born in 1925 in Kassel, Germany, Bernhardt studied at Frankfurt University and in 1955 attained his doctorate in jurisprudence, specializing in comparative public and international law. He began working at the Max Planck Institute in 1954, initially as a research assistant, later as a research fellow, and was appointed its Director in 1970.

Many of the social problems with which he is concerned are a direct result of scientific progress. He believes that scientists, politicians and lawyers all have a responsibility towards society in this respect. "Although science is a major contributor to human well-being," says Bernhardt, "it also has the disadvantage of being a potential source of harmful developments. Science is as much part of the problem as part of the solution, because discoveries are rarely made with an opportunity of seeing future implications for society."

The advancement of natural and applied sciences has led to discoveries which have been taken up by industry for commercial gain, and whose implications are only considered by politicians at a later stage. Bernhardt cites many topical examples such as the development of genetic mapping, the application of gene splicing in biotechnology, depletion of the ozone layer, deforestation in South America and proliferation of nuclear energy generators. "These developments are familiar to greater numbers of people than ever before," he continues, "because of the popularization of science in the media and our heightened recognition of the quality of life."

Dealing with complexities

It is not a simple matter to cope with these social issues, and like
anyone else, lawyers have to contribute as individuals as well as in a
professional capacity. According to Bernhardt, changes now occur
so quickly and are often of such a fundamental nature that it is no
longer enough to simply analyse existing law. "Lawyers must also
look for the deficiencies in existing legal systems," he maintains.
"We should find a balance between consistency and change that
will hopefully lead to progress, using an intricate process of response
and analysis. Nowadays, problems are particularly caused by the
fact that it's often impossible to clearly see potential dangers in
areas of rapidly moving industrial developments, such as nuclear
engineering and bio-engineering. It is also difficult to determine
what should be done to prevent these dangers becoming reality."

Certain developments, for instance transportation of chemical
wastes, terrorism and drugs trading, have an immediate impact on
society in contrast to the more distant nature of issues such as
nuclear energy generation and depletion of the ozone layer. "But
both types of problem have an international character," says
Bernhardt, "and the legal consequences of worldwide activities
becoming more international are very complex. Politicians should
be leaders, with lawyers taking the role of servant."

Another major problem confronting lawyers is that it is often
difficult to find an objective authority for the issues at stake. "Take
the recent example of steroid hormones," says Bernhardt. "Politicians have been accused of not using the law solely to take care of
health, even though it was designed for that purpose. It has instead
been suggested that they are abusing the law for economic reasons,
in order to please the agricultural pressure groups." In Bernhardt's
opinion, it is impossible for politicians, or lawyers, to give an
absolute answer to this particular question of whether certain
hormones are dangerous to health. He sees it as equally impossible
to determine whether the law is used merely as a competitive ploy
by European against North Americans, or vice versa.

"Society at large," says Bernhardt, "needs independent-minded
scientists with no allegiance to industry, commerce, politics, pressure
groups, or anyone with direct or indirect vested interests in a

particular case. Not only is independent science vital for society, but it is also essential if lawyers are to function effectively."

The case of steroid hormones, together with many other recent or distant examples, proves that science can not operate independently without a controlling mechanism. Bernhardt considers that this is true of all critical social functions which need to perform independently, including private enterprise, politics and law. "There is no stereotype available to create an ideal balance between these activities," he points out. "But even so, lawyers and independent scientists must become increasingly capable of complementing each other when dealing with issues on an international scale."

Furthermore, he recognizes that lawyers depend on politicians, entrepreneurs, scientists, and the public at large, just as much as each of these groups depends on the other. According to Bernhardt, interdependence is intrinsic to the general quality of life. He believes that "quality of life means quality of interdependence, it means reaching a consensus of relations."

The role of the European Court

His responsibilities as judge at the European Court of Human Rights in Strasbourg enable Bernhardt to directly witness and take part in one form of cooperation in practice. The Court functions within the Human Rights programme of the Council of Europe and now comprises 23 judges, one for each member state. These member states are parties to the European Convention for the Protection of Human Rights (1950), which stipulates that there should be both a European Commission and a European Court of Human Rights.

The Commission examines complaints made either by a contracting party or, in certain cases, a non-governmental organization or group of individuals. If the Commission decides to admit the application, it places itself at the disposal of the parties in order to attempt reaching an amicable settlement. If no settlement is achieved, a report is sent to the Council of Europe's Committee of Ministers. The Committee, or the Court whenever applicable, decides whether or not a violation has taken place.

Since 1955, more than 15,000 human rights applications have been lodged. The Committee of Ministers is responsible for implementing the Court's decisions, although few sanctions are enforced because member states usually follow the rulings if they lose a case. "The system is based on a clear intention of cooperation," Bernhardt explains. "I'm happy to say that, in general, our jurisdiction is accepted and implemented. This may take considerable time in some cases, which is understandable because the legislation of the member state involved may have to be changed. Procedures for changing legislation are usually time-consuming, especially when alterations are imposed by external instead of internal pressure."

Bernhardt says that examples of a state not following the decision of the Court are few, even when the time factor is taken into account. "Public opinion plays an important role here, and pressure from voters and public awareness are essential," he explains. "But politicians *have* to act, in any case. If you look at such subjects as population growth, food shortages, or depletion of natural resources, only a global approach can provide relief."

"You often come across scepticism about whether mankind is capable of coping with the scale and complexities of such multidisciplinary and multinational issues, but I'm optimistic about the potential for making progress." Bernhardt's optimism is based on his experience in law, as a scientist and as an experienced judge of an international court, all of which enable him to be convinced of the direction which should be followed.

Interaction between states

Bernhardt does not believe that the ultimate goal should be a World State with one legislation and a single government, and finds that both an impossible and unattractive situation to strive for. "The interdependence of individual states is not the same among states."

Indeed, Bernhardt thinks that even the idea of a single World State is irrelevant, so long as such an option remains unrealistic. "There are nearly two hundred states worldwide. You often hear about differences between European and North American legal systems, but as a lawyer, I'm also aware of the serious legislation differences which have to be dealt with *within* the West. In fact, there are more differences between the UK and continental systems than between Europe and America. The UK doesn't have a written constitution, whereas both Europe and the United States do."

"Case law and common law might also have played a greater role in Britain than in America," adds Bernhardt, "but the European Community is bringing the UK and Europe closer together. I don't see such great differences between European countries, North American and Latin American states because the latter usually inherited continental systems."

Bernhardt indicates, however, that not only coordination but also integration might be far more difficult to achieve in the case of African and Asian systems. "We have African students here in the Max Planck Institute," he says. "It's very difficult for us to understand their systems, partly because we don't know enough about the details. It's very hard to describe these systems properly, but understanding can only begin when we have sufficient facts."

In the case of China, Bernhardt has learnt from Chinese colleagues that they believe it will take some forty of fifty years before standards of living and levels of knowledge and understanding are sufficiently similar for coordination and cooperation to begin. Only when this is achieved can the process of assimilation, exchange and integration start.

The law follows, never leads

Bernhardt then moves on to consider the irritation which is increasingly being directed at legal systems. Claims are frequently made that the law is misused and that instead of safeguarding individual rights, it creates barriers which obstruct efforts to make society work effectively. Criticisms are levelled at costly, complex and lengthy court procedures which seem to victimize complainants and benefit only lawyers, and whose outcomes appear to favour the wealthiest party with the best legal counsel.

Bernhardt reacts firmly to such cynicism. "I wouldn't use expressions such as 'misuse of law'. In my definition, that would mean jeopardizing the traditional legal system when no institutional alternatives are available. There are always deficiencies in any system, including the law. But nowadays it is often criticized for being too tolerant and respectful to individuals, as alienating them by a

seemingly endless series of appeal procedures in litigation. Once
again, the jurist is simply applying the law."

In Bernhardt's opinion, conflicts of interest between individuals
and society can not be pacified by a judge in court. He believes that
this is the role of politicians, economists and scientists, because
lawyers can only specify in what way legal systems can expand or
adjust. "Legal systems must be continuously improved," he says,
"but even though inadequacies in the functioning of society force
urgent change, lawyers are not equipped or entitled to do anything
other than practise the law."

"Neither too much nor too little should be expected from codifi-
cation, legislation or the courts," Bernhardt suggests. "We are
living in a society that is gradually distancing itself from rigid
ideologies, totalitarian regimes and dictatorial structures or systems.
That's why secularization took place, and why the communists
recently lost control in Eastern Europe and the Soviet Union. We
shouldn't expect lawyers to reside in remote courts and impose
truth, as such a society would make them agents of absolute justice.
Politicians rely on elections and entrepreneurs on market response,
but as I said earlier, the law can only follow but never lead."

Mari Pijnenborg

Rudolf Bernhardt is crystal clear: lawyers will not take the lead in creating a framework for the quickly changing needs of a modern high-tech world. But who will lead us? "I can imagine scientists, philosophers and other people with specialist knowledge taking the lead in global control," says Roger Penrose, an Oxford Professor of Mathematics and author of the book on Artificial Intelligence 'The Emperor's New Mind' (1989). "Scientists could start informally," he says, "initiating something like the Pugwash Conferences, only broader as they were limited to nuclear weapons. Nowadays, we have many other serious current topics than the nuclear threat alone."

"You can always ask yourself whether even a man like Gorbachev was influenced at all in his opinion on arms reduction policies by the Pugwash Conferences. Conversely though, governments do exert a serious influence on science. And in that respect, the unilateral emphasis on economic benefits focused on national interests is far too big. The intrinsic value of science is not recognized enough. This should be stipulated more, together with emphasis on a more global approach."

ROGER PENROSE, the mathematician

The black hole
of consciousness

Professor Penrose points to the extremely difficult problem of how
we should control the social implications of science. "Global organization would be needed as it is beyond the reach of individuals or
even a simple country. I may be a little Utopian in this, but such an
organization should have power over countries, companies, and
governments. Good communication has an essential part to play,
and needs to have control too. Everyday, you see examples of how
quickly misunderstandings about scientific findings can occur. The
other day one of the newspaper headlines screamed out: 'Milk in
microwaves causes brain damage', a gross misinterpretation of some
research findings. To avoid unnecessary scares, science communication needs a certain authority to place control on it." Central to
science and control, of course, is the working of the human mind. A
television programme by the BBC (British Broadcasting Corporation) about the likelihood of omnipotent computers eventually
replacing the human mind, was the final impetus needed for
Penrose to write his book 'The Emperor's New Mind'. In the book,
aimed at a general readership, he strongly opposes the possibility of
computers in their present configuration taking over human thinking.

Penrose gained his scientific reputation for his work in areas
other than Artificial Intelligence. He is perhaps best known for his
contribution to a theory explaining the presence of Black Holes in
the Universe, formulated together with Cambridge Professor
Stephen Hawking in 1965. In 1974, Penrose was also responsible for
discovering pairs of mathematical shapes, the so-called 'Penrose

Tiles', which unexpectedly turned out to underlie, in a three-dimensional projection, a strange new kind of matter called 'quasicrystals'.

The mystery of human thinking

In his book, Penrose states that human thinking is a process involving more than mere computation and argues that although high-speed computers will have a considerable impact on society, overestimation of Artificial Intelligence is too farfetched and may even be dangerous.

"My field is not really computers, but the more I experienced how widespread the belief is that computers will soon be able to take over human thinking, the more I thought that I must do something," says Penrose. "What stimulated me most to write about Artificial Intelligence was the feeling that people are missing the point. I was obviously not the first to oppose the possibility of 'conscious' computers. John Searle, for example, of the University of California in Berkeley, stresses that computation in itself does not evoke consciousness, or mental phenomena such as awareness, pain, hope or understanding. Nevertheless, his viewpoint does imply that a sufficiently sophisticated and complicated computer program could eventually achieve a successful simulation of the outward action of the brain."

"My argument goes further than this. I say that even apart from the lack of 'consciousness' in computer operations, there is more to human thinking. There is a necessary ingredient that is essentially of a non-computational nature, which until now, no-one seems to have grasped. I have always believed that thinking is non-computational, and even though the field of computers is an outside issue for me, I have always believed in the possibility of non-computability. So I decided to provide evidence for that in a general book."

His 460-page book examines what physics and mathematics can tell us about how the mind works. It ranges over a wide number of topics such as relativity theory, quantum mechanics, cosmology as well as many philosophical issues. Penrose points out that the nature of consciousness and its impact on the working of the mind still bears many mysterious elements. "Without a 'theory of consciousness', contemporary scientific knowledge is still too limited for it to be able to provide a description of the mind's work," states Penrose.

A remarkable parallel can be seen to exist between the way in which Penrose is concerned about public attitudes towards the role of computers nowadays, and the way in which his father, Professor Lionel Penrose, opposed the belief prevalent during the rise of Nazi-Germany that human genetics could improve mankind. When his father was appointed to the Galton Chair for Eugenics in London in 1945, he insisted on changing the term. "My father was emotionally against the idea," says Penrose, "from a deep concern about the social implications of such science. But he also acknowledged that scientific knowledge of genetics was nowhere close to really succeeding in this. Instead, the closest applicable solutions were to be found in the diagnosis and cure of some genetic defects. So that is where he laid the emphasis of his work."

Consciousness theory

Does Penrose believe in the possibility of a 'Grand Theory of Consciousness', if he believes that it is a prerequisite for imitating human thought? "I'm an optimist about what science can achieve," Penrose answers. "I see no reason why we shouldn't come to such a theory – although I agree, we are still a long way from it. There is a close analogy in our understanding of Black Holes in the Universe. If Einstein's relativity theory holds all the way down, then Black Holes also represent something at their cores where the laws of physics no longer apply. We lack a theory for explaining these singularities. There is actually a big gap in present understanding between classical physics, explaining the 'big', and quantum mechanics explaining the 'tiny'. It is necessary to first bridge this gap in order to understand the essence of the laws of physics better. That won't put everything into place immediately, but it is the first essential step to come to a scientific theory of consciousness."

"Right now, a common presumption is that if something acts according to the laws of physics, then it is computational. I believe this is false – the non-computational element is essential. Even if we

are still a long way from it, this doesn't prevent us from commenting on such a non-existing theory. My claim is that a necessary
ingredient for this theory is that it is non-computable and that this
will allow a physical description of consciousness that is non-computable. How does thinking work? Mathematicians clearly recognize a
kind of reasoning that can not be formalized, which touches on the
essence of understanding and meaning. Gödel showed that just
manipulating symbols, as in the formalist definition of mathematics,
was limited (Gödel's Theorem, 1931). It would never encapsulate
mathematical truth. Instead, the essence is understanding what
understanding is, rather than manipulating symbols. Not everything
is computable in mathematics. Otherwise the 'Ah!' of awareness
would clearly be missing."

In 'The Emperor's New Mind' Penrose vividly illustrates how
inspiration can occur, which he describes as 'occasional flashes of
new insight'. His famous theorem on the singularities in Black
Holes, for example, occurred to him while crossing a street. Later in
the day, this caused 'an odd feeling of elation'. Going through his
mind for a reason for this feeling, the particular thought when
crossing the street came back.

Similarly, he had absolutely no sense of the future significance
which the Penrose Tiles would have for materials science when he
first discovered them. In an almost playful way, Penrose had set
himself the goal of finding a non-repetitive way of tiling a plane
with identical shapes. About the actual outcome Penrose still says:
"It is hard to express in words, but I knew there was something
more. I was struck by the beauty of it. Visually, but also by the
beauty of the concept." He goes on to explain that much of his
thinking is visual and states that rigorous, verbal argument is,
therefore, usually the last step when forming a concept.

"If we come close to a theory of consciousness, our view of the
world might of course be changed by it. You can't tell how, or what
might be constructed then. The social implications of the computer
– at least as it exists in its present form – will remain. With thought
comprising a non-computational element, computers can never do
what we human beings can. This doesn't mean that they can not
partially imitate us. So, the strong point about Artificial Intelligence
is that computers could take over, but without the consciousness.
How dangerous is that? One sometimes finds people who are more

like computers, and sometimes these people also get too much power. So what is the difference, you might argue."

Penrose doubts that a theory of consciousness will be the ultimate contribution of science to society, or that science will then be complete. "Every discovery always raises its own problems," he explains. "Einstein, for example, resolved many puzzles by his theory of relativity, but it raised many new questions. The lack of a theory for the singularities in Black Holes is just one of them."

Controlled spread of knowledge

"If we had a reasonable grip on consciousness," Penrose continues, "I am convinced that new things, new opportunities would emerge. But there is also the possibility of abuse. Who will manipulate consciousness? Finding an explanatory theory for consciousness or the abuse of it, raises other questions. Is it a good thing in itself? Is knowledge a good thing for its own sake?"

"The danger of an unequal spread of knowledge in society is of course not a new phenomenon. But I really think that some kind of effort is needed on a global scale to keep the process under control. I am a pessimist if the 'small is good' principle is applied here. Such things should not be left to individuals."

"Let us look at how computers affect society," Penrose goes on. "The great advantage of the computer is that it has led us in many new directions, by enabling us to cope with more data and larger numbers. In science, for example, we are now able to work out the consequences of many theories very much more effectively. The negative aspect is that computers give much of the work a certain authority which is not deserved. A prevalent attitude nowadays is that if it's done by a computer, it must be right. The computer though is only a tool, albeit a very valuable one. Computers enable more to be done, but you still need the understanding, and *that* it doesn't provide."

"Computers do allow people to cope with more information and are strongly intertwined with communications. Recent events in Eastern Europe and South Africa, for instance, show that people no longer live in isolation. Everyone can now see what happens in the world, so in that sense they all affect each other. In the regions I've just mentioned, it clearly had a positive effect."

"On the other hand, people just let it happen. We don't really seem to make use of all the available information. We could ask about cause and effects. To what extent is it that computers are changing society? I would say it is as much advances in communication and other technological developments. Coping with these new phenomena poses difficult problems. Is there an inherent instability in scientific progress, as can be seen in the present developments?"

"Because worldwide information flows mean that the world is increasingly looked at as a unity, planning and understanding become essential. The main flaw at present is that people don't have global plans. They all work for their own interests and it takes time to realise what the effects of an individual action are."

Penrose stipulates that, once again, scientists, philosophers and other people with specialist knowledge have a role to play by carefully considering the societal implications of their work. But the same Penrose, of which the imaginative mind came up with far-reaching new theories and ideas of which implications were impossible to foresee, recognizes that this is not always possible.

Some discoveries lead their own life. "What has always been most cited from my publications is a certain side-issue that I obliquely covered in my first year of research," Penrose illustrates. "It was a generalized inverse for matrices and I have no idea how people apply it nowadays, but I do know that a whole conference was dedicated to it and to related ideas several years ago here in Oxford. The same happened with the quasicrystals, which seem closely related to the Penrose Tiles. They may have special electrical properties which, however interesting, I can not explain adequately as I am not in that field. These events prove the difficulty of being unable to foresee where the discovery is ultimately going."

The value of knowledge

Stipulating, once again, the intrinsic value of science, Penrose says that he had more than one motive when writing his book. "I also wanted to explain to the general public what I had myself learned about the world in all these years of rather esoteric science. I wanted to describe the intrinsic value which science has, comparable to the worth of artistic possessions. Science is not just necessary for social progress, but for civilization itself. This value is quite independent of any social implications."

"When I was younger I tended to cut myself off from the outside world. It was important for me to go my own way. This has changed. When one gets older, one has more commitments to the outside world. I sometimes regret not having more time to think on my own, but equally, one becomes more aware of the importance of relating to the outside world. My book is a very concrete example. I believe people benefit from understanding the world, if only for the joy of it, for its own sake."

"Aside from this is the fact that education is important to cope with the world. Academics, big companies and governments know more about the world, but they should not have a lead over the rest of mankind. Ordinary people should get the information, they should be provided with the available answers to the basic question: 'why?'."

Eefke Smit

The intrinsic value of science is a commodity itself, states Roger Penrose. He believes that people benefit from understanding the world, if only for the joy of it, for its own sake. But then, people have to read more, stresses the Swedish physicist Professor Kai Siegbahn, who lives and works in Uppsala. "Young people aiming at doctoral degrees, often don't read beyond what is essential for their theses," he says. "They don't attend seminars on topics not specifically related to their particular field of study."

"To get surprising ideas, extensive reading is essential. I enjoy reading. I am a member of several editorial boards, frequent the library of our institute and receive eighteen journals at home. I look at the titles to find something useful, even in the advertisements in the journals. Reading is a very important part of my job. Young people don't seem to understand the importance of it. You can't just sit there looking at your apparatus. You have to read a lot. Then you get new ideas."

KAI SIEGBAHN, the physicist

The accelerating advances

Extensive reading suits Siegbahn, coming from 'a sort of professor's family'. His father, Karl Manne Georg Siegbahn, was awarded the Nobel Prize for physics in 1924 for his discoveries and investigations in X-ray spectroscopy. Kai Siegbahn was honoured in 1981 with the Nobel Prize for physics for his development of electron spectroscopy. It is no wonder that the Swedish Professor is intrigued by the enormous changes in science and society that are taking place at a breathtaking pace. In an essay entitled 'The Scientist in a Changing World', Siegbahn observes that the role of the scientist in society has changed drastically during the last two or three decades.

"We may recall with nostalgia the time when science was in the hands of a small number of individuals with little or no contact with society in general. The research was done in small laboratories by the scientist himself, maybe with the help of a devoted assistant. The great change in scientific life occurred when it was realized during the Second World War that large scale scientific team work could supply new alternative powerful military tools."

Yet Siegbahn does not really look back with nostalgia. Enthusiastically, he sums up the advancement of science since the early forties when he began his scientific career. "We have witnessed the discovery of the liberation of nuclear energy, the invention of radar, television, the transistor and semiconductor techniques, the first computer or as it was first called, the 'mathematics machine', and the complete 'datorization' of all administrative activities and of techniques and communications." He adds to the list of scientific achievements: micro-electronics, the first rocket into space, the introduction of antibiotics, the discoveries made in genetics and molecular biology, biotechnology on an industrial level and the discovery that the ultimate primitive form of matter consists of quarks and not of atomic nuclei. These scientific and technical breakthroughs, which have led to fundamental changes in society, may be followed by equally new and unexpected discoveries in the coming decades, Siegbahn anticipates. And these may change the conditions of life in the same rapid pace. "It affects directly our own generation and that of our children."

International scientific cooperation

The scientific community and political leaders should cooperate, Siegbahn is quick to point out, to make sure that the scientific and technological evolution benefits mankind. This cooperation should be established worldwide, not only in the industrialized world, he continues. "Scientific work and collaboration across national and ideological borders is an efficient means of transferring knowledge, creating new channels for mutual understanding and respect, and initiating appropriate industrial and political developments."

International cooperation in science is of special interest to Siegbahn. All scientists have an international responsibility and should act accordingly, even on a small scale in their day-to-day activities, he is convinced. Scientists from all over the world have always been invited to participate in Siegbahn's own research group. In his home, about a ten-minute walk from the laboratory where he continued working after his retirement, Siegbahn explains: "We started by organizing what we called the International Seminar in Physics for about twenty physicists from various countries throughout the world. The idea was to give them training and have them contribute to international scientific work. The seminars would boost their self-consciousness and self-appreciation."

"Of course these people were rather lonely at night, because we had organized their stay in a bad way. They should not only have a place to work, but also a place to sit in the evenings. So an architect rebuilt a storeroom of our institute into an extremely nice club-room where they could have a fire in the evenings and make their tea."

"The spread of scientific knowledge is essential, also to poorer countries, which are not so well-developed industrially. We should not extend the technological gap between the developed and developing world. Every country must be able to build up its technology and scientific knowledge, in order to raise people to a higher level, so that they can produce their own teachers."

"In recent years, collaboration between scientists in the industrialized and developing countries has grown. The technology gap, however, is growing because of the high speed of developments in the industrialized countries."

Professor Siegbahn is also involved in an organization called The World Laboratory. This decentralized organization has projects in China, India, Iran, various countries of the Mediterranean, South America and elsewhere. A small central office administers the funds and scholarships and coordinates activities. "My laboratory, which is quite modest, belongs to the organization," he explains. "We receive guest scholarships from The World Laboratory and scientists stay here for a year or two after which they return to their own countries."

"Often the resources and conditions they meet at home do not match those they had when working in laboratories in industrial countries. So the cooperation must continue when the scientists return home. Maintaining contact is a very important part of the whole scheme. Usually we visit them and try to help financially as well as to coordinate the work in that country."

Father of electron spectroscopy

Kai Manne Börje Siegbahn was born in Lund in 1918. He went to school in Uppsala, and in 1936 entered the University of Uppsala to study physics, mathematics and chemistry. "After the usual degrees taken by most other young people at the time, I chose nuclear physics as the subject for my doctoral thesis. This was an exciting new field with extremely interesting prospects. This turned out to be a good guess because in the following decades nuclear science developed tremendously. At that time nobody had a clear picture of the structure of the nucleus. I was very interested to see if one could do some sort of spectroscopic work, not the usual optical spectroscopic work, but based on radioactive radiation. So I set about designing instruments and making experimental arrangements to investigate this radiation. That was the work for my doctorate, and afterwards."

"At the end of the forties, the international scientific community got a new view of the structure of the nucleus which was found to have some sort of shell structure. This was not expected at all from earlier experiments, which had given no indication of this shell structure so well known in atomic physics. That the electrons circulating the nucleus were confined to various shells, was the basis of optical and X-ray spectroscopy, which so far had described the atoms and the molecules. But for the nucleus, which was compressed to a volume many, many orders of magnitude smaller than the atomic electron shells, any sort of shell structure in that very tiny volume was thought to be very unlikely."

"This was, however, the case. It made nuclear spectroscopy, which I was devoting my work to, much more meaningful. Now the structure of nuclei could be studied in a systematic way, and clear properties could be attributed to the various energy levels in a nucleus. I worked hard for several years in order to try to understand what I had to do, experimentally, to reach my goal. I wanted to get some sort of spectroscopy applied to solids, molecules and atoms, and using X-rays as a primary source to expel electrons from these materials."

"We were getting very new results which showed that a new type of surface electron spectroscopy could be developed. Every day of our work my students and I found something new," he explains. Yet the most useful part of their discovery, Siegbahn and his collaborators first perceived as a troubling distortion of the results. "We observed that the electron spectra that we obtained from a certain element depended on the chemical condition of that element. We found what we call the chemical shifts. In the beginning we didn't like this. We were physicists and wanted to study systematically the behaviour of elements. It was a problem as now we had to be careful that the substance we used was not oxidized or changed in some other way."

"Eventually we realized that this was precisely the beauty of the whole method. We could not only observe the elemental composition of the surface, but also the chemical conditions on the surface. This has been of great importance for surface studies."

Nowadays, electron spectroscopy is routinely used on an extensive scale to study corrosion, for example. But after the first discoveries in the laboratory, there was still a long way to go before industrial application would be possible. "This had to be developed gradually from all points of view. I got my first results in about 1957. Some years later, in the mid-sixties, we had collected so much data and so many spectra that I decided we should write a book together. To push the work, we worked in shifts, day and night, ever day of the week, for a year until we had collected enough data to write the book."

"We had been publishing continuously of course, but only in the rather obscure journal of the Swedish Academy of Sciences. The journal was not widely distributed, being used mainly to publish doctoral theses. Our material was in the annals of the Academy, and there it stayed for some time. We didn't realize its importance. The usual procedure nowadays is to send drafts to interested colleagues for comment and then after further complementary work to submit the manuscript to an international scientific journal."

The book entitled 'Electron Spectroscopy for Chemical Analysis', or ECSA, was published in 1967. It describes in detail the new spectroscopy, setting out the advantages and potential. Since then, Siegbahn's work has focused on electron spectroscopy and its applications. While on sabbatical leave in Berkeley, California, he helped the American company Hewlett Packard design a technically well-developed instrument for commercial production.

In 1981 Siegbahn was awarded a Nobel Prize for his work on electron spectroscopy. He shared the prize with the American physicists Nicolaas Bloembergen and Arthur L. Schawlow, who got the other half of the prize for their pioneer work in laser spectroscopy. The University of Uppsala let Siegbahn keep his laboratory after his formal retirement so that he could continue his research on the combination of laser with electron spectroscopy.

Science and society

Siegbahn is a scientist in heart and soul. He believes in "a scientific approach based on idealism and technical and ethical knowledge." Science, as he sees it, has an important role to play in solving the major world problems. It should provide society with the basic understanding of complicated issues, such as the catastrophic increase of the world population, the precarious food situation, environmental problems, and the growing technology gap between the industrialized and the developing countries. "All scientists worldwide have a prime duty," according to Siegbahn, "to contribute their knowledge to solving these problems. Continuous interaction between scientists and politicians is needed to put this into practice."

"A scientist is just a scientist and is not necessarily better at politics than anyone else. But he is, of course, accustomed to untangling problems in a scientific way. International in attitude, scientists travel around visiting laboratories, giving lectures, attending seminars and meeting people. CERN in Geneva, for example, has been instrumental in bringing together people from the Soviet Union, the United States and the countries of Europe. The World Laboratory is another example."

"How to use science is a political question," he continues. "But scientists often observe problems before politicians do. Long ago scientists warned politicians of the grave environmental consequences of industrial development. Now political decisions have to be made on how to face these problems. Once these decisions are made, scientists will be involved again. We have to do a lot of research to find out what we have to do to protect the environment. We need to know the effects of certain measures, to prevent that more harm is done in trying to repair the damage. The scientific approach can not be side-stepped."

Scientists themselves, according to Siegbahn, are becoming increasingly interested in the relevance of their activities for society. And society at large is gaining better insight into what happens in science, thanks to all sorts of popularization. "To the great satisfaction of scientists, the press reports more on science. I often read about great and small scientific discoveries in the newspapers before I read about them anywhere else."

"Nevertheless, many scientists are a little afraid to approach the mass media, because they never know what the result will be. They may be quoted out of context, or in a sensational way, for example. The journalist wants to make big news, the scientist a big career, and sometimes this makes the communication between them difficult. In the case of cold nuclear fusion, which seemed to be a sensation, the importance of the experiment was overestimated."

"In science it is typical that you can't forecast precisely the importance of a discovery. The grand old man of nuclear physics,

Lord Rutherford, who made all the important discoveries in the field, gave a lecture to the Royal Society in which he said that any suggestion that nuclear energy could have a useful purpose, was 'blue moonshine', just fantasy. A year later nuclear fission was discovered. That shows that in science you can't see around the corner. There are always surprises."

Juurd Eijsvoogel

A scientist in heart and soul, Siegbahn believes in a scientific approach based on idealism, as most Western scientists do. He thinks it "a prime duty of all scientists in the world" to contribute their knowledge to solving global problems.

Japanese are not well accustomed to this approach, suggests Professor Shigeo Minowa, Director of the Institute of International Business and Management at the Kanagawa University in Tokyo.

SHIGEO MINOWA, the publisher

The communication deficit

"In Japan there is no strong concept in scientific practice as to what does or does not contribute to the well being of the whole world. Values of 'good' and 'bad' as in Western society do not exist. Here, everything is allowed in science, as long as it is aimed at catching up with the rest."

Minowa believes that the vision of Japanese scientists and students is too limited by their own cultural attitude. He stresses that this is "in itself a peculiarity of the Japanese people." He says: "If you combine this with the Japanese trait to concentrate vigorously on details without religious, ideological or philosophical constraints, then the problem of 'science beyond control' is bigger in Japan than anywhere else. The interest of scientists is unlimited, and they do not look at the social implications of their work."

So, one of the things that Professor Shigeo Minowa constantly tries to teach the MBA-students at his institute, is the duty of Japanese to make themselves understood by the rest of the world. He encourages them to learn to communicate better with people from other cultures and to try to understand other ways of life, other ways of thinking. "That is not so easy," he says. "In general, the Japanese are inexperienced in dealing with foreign countries. Japan is good in information collection, but not in communication."

Minowa, who has been a science publisher for most of his career, and as an economist has long been studying quantitative trends in the relation between the economic development of different countries and their relative importance in communication or publishing industries, expresses personally a keen interest in a well balanced, healthy position for Japan in the world – in the economy, in science and in politics. But he has his concerns on each of the three.

"From an international perspective, Japan is first class in economics, but at best second class in handling politics, and Japan is definitely in the lower ranks regarding communication and publishing," he begins. "As science is directly dependent on good communication, this bears its implications on Japan's position in world science."

He smiles teasingly when reminded of the fact that Japan counts more telephones, more televisions, more newspapers, faxes and copying machines per capita than any other country in the world. "These are no more than means for our industry to sell more and to sell better," he says. "It is not per se a sign of more communication."

"The per capita ratio consumption of paper is extremely high in Japan in relation to the population – but just look how much of our newspapers is used for advertising! A so homogeneous and densely populated country as Japan is a paradise for marketing people."

Import of information, export of goods

As the former head of the Scientific Information Resources of the United Nations' University in Tokyo (1975–1985), Minowa is committed to better the international relations between nations.

He is convinced that Japan should strive to balance the present unevenness in its contribution to worldwide communication. "In the present situation, Japan will not easily reach a trade surplus in the publishing industry," he muses. "Japan buys more information than it sells and paradoxically that relates directly to its exporting cult in most other sectors. If you want to successfully export industrial goods, you have to start with collecting information about the markets you want to enter."

"Most things done by Japan have their roots in a commercial interest, not in a certain philosophy or in deep thoughts. This counts for science as well as for trade. In science, Japanese seek information in order to catch up with the rest and to apply scientific findings for better products. Japanese tradesmen want information on certain countries, but care very little for the sentiments of local people in the rest of the world. They just want to sell. In that sense the present economical success is caused by short term naïvety."

"Japan would love to go on living its own isolated life. In that sense our international business people are two-faced. They may look Westernized, they may sell in the rest of the world, they may create more economic interdependence, but they hate to adapt their thinking to other cultures. In their hearts they remain very traditional and Japanese. There is a danger there. For just as little do they make the outside world understand what drives Japan and thus threaten the possibilities for Japan to live harmoniously with the rest of the world."

"This is also reflected in the way science is executed in Japan. Japanese scientists are not interested in the impact of their achievements elsewhere."

A problem of control

Minowa even confesses to be 'desperate' if he sees how science in Japan works. "Naïveté rules," he says. "There is very little interest in the relation between science development and the values of human life. And if you should ask me, honestly, I would not know how to solve this. So many scientists are convinced that if only enough progress is made in science, they will be able to control everything. But it is my belief that something may go terribly wrong because suddenly there is no mechanism to control it any longer."

"We are nowadays a 'science society'. Everything relates to it: education, food, culture, industry. Science and technology is at present strongly interconnected to the economy and military affairs. The problem is that most people do not understand what is happening in the black box of science – the implications on society of scientific progress go beyond our containment."

"What we do need is a more critical stance towards science, maybe more critical communication. The involvement in science issues should stretch from governments and politics to industry and may not exclude the average citizen."

"Citizens have problems understanding what science does, so they need good and sound information. Wisdom is the key to control. Publishers have a role to play there."

This brings Minowa to the field in which he – as an economist – has focused so much attention: the relationship between economic development, the publishing industry and society.

Education affects demand for information

Minowa has described in quantitative terms the trends within each
developing society how publishing activities suddenly undergo a
take-off and how a steep rise in the number of titles published
emerges.

In Europe, around 1780, Germany was the first country to
exhibit this trend. Most other nations came much later: the UK in
1820, France around 1830, the US around 1850 and Japan around
1870. Minowa: "Before these dates, there were no more than 700
new titles per year in each of these countries. After the 'boom' in
the years mentioned, the level rose to many thousands per year."

Surprisingly, this 'publishing trend' neither correlates to the
industrial development of a country, nor to its achievements in
science. The sole explanatory factor however lies in the quality of
general education.

Minowa: "The better the education system, the more people
become receptive for information. The Prussians in Germany, for
example, had established a well-developed school regime at the end
of the eighteenth century, while the UK lagged behind in education
those days."

"You see the same thing in Japan, which had been oppressed
and isolated until the second half of the nineteenth century. After
that, things quickly caught up: the nation began to develop at an
increasingly high pace. People became more and better educated
and a strong demand for information developed. And education is
in itself a strong impetus for generating information."

"Publishing is important for educating people," Minowa concludes. "And that is where science and society come together. For
scientific progress, information is indispensable. To involve the
public at large in the issue where science takes society, education
and communication is crucial."

"In a society that is getting more and more 'sciencified', the role
for publishers is bringing information at such a level that people are
able to assess what science is capable of."

Eefke Smit

Erich Bloch, the director of America's National Science Foundation (NSF), points at the danger that some people who do not understand the quick developments in science – 'the technically illiterate', he calls them – will not be able to form their own opinion. That they will simply look at science as the way, the only way to solve problems. Science being a sort of almighty God who rules the world. "That's a problem, because science is not almighty," Bloch says. Shigeo Minowa stresses the need for a more critical attitude towards science. "Citizens have problems to understand what science does, so they need good communication. Wisdom is the key to control," Minowa says. Concerning this problem, Minowa thinks publishers have a special role to play. So does John Maddox, Editor of the influential science journal 'Nature'.

JOHN MADDOX, the editor

The public digestion

"There has been a spectacular success of science and subsequent growth of a belief that anything is achievable, be it development of drugs which are entirely free of side-effects or the possibility of industry operating without a trace of pollution. Expectations, in other words, have become quite unrealistic," Maddox says. "The problem is worsened by governments that tend to tell voters that risks can be totally eradicated; for example, when there is a scare about the side-effects of some drug. It's a weakness of governments to go along with the idea that everything can be cured."

He cites the recent banning of steroid hormones in animal husbandry by the European Community's Council of Ministers as 'a big scandal'. The action was taken against the advice of the EC's own expert committee and in contradiction to the judgements of the WHO (World Health Organization) and the US Food and Drug Administration. He insists that the European press was partly to blame. "They accepted uncritically that an irrational decision was nevertheless convenient to those ministers and commissioners concerned with agricultural policy as a means of keeping down beef and milk surpluses," he explains. "The reporting of that event was cynical. The European press doesn't yet seem to have grasped the importance of contributing towards formation of responsible European governments."

Speaking out on scientific issues

As the Editor of 'Nature', a position he held between 1966 and 1973 and now again since 1980, Maddox has his own role to play in responsible reporting. He is a stout defender of scientific enterprise, which he describes as "the essentially communal activity of science, which has the quality of adventure and now involves millions of people trying to understand how the world ticks and doing so on the whole with an unusual degree of camaraderie." His journal he sees as "a vehicle both helping that process to work and influencing how it works." Under his current editorship 'Nature' has doubled its circulation and now sells 33,530 copies weekly.

After a career in theoretical physics at the University of Manchester, Maddox was science correspondent for the 'Manchester Guardian' newspaper from 1955 to 1964 before moving on to become editor of 'Nature' the first time. In the early seventies, he was Managing Director of Macmillan Journals Ltd. and then became Chairman of Maddox Editorial Ltd., whose publications included 'European Gazette'. Over the years, he has also been an affiliate of the Rockefeller Institute in New York, Director of the Nuffield Foundation, London, and Public Interest Representative on the UK Genetic Manipulation Advisory Group. Maddox has written several books including 'The Spread of Nuclear Weapons' with Leonard Beaton (1962), 'The Doomsday Syndrome' (1972) and 'Beyond the Energy Crisis' (1975).

Coming back to the steroid hormones scandal, he says that "the incident also demonstrates that a healthy society needs reliable scientific information at all levels. The citizen and voter in particular need a much more thorough understanding of how governments are both using and in many cases misusing science. This means that more general reporters need to come to grips with scientific problems."

"A specific initiative I would personally like to see is the launching of a high-quality technical newspaper aimed at the general public, which would cover important issues of the day, such as growth-promoting hormones and the greenhouse effect. This would help us all to deal with the pressing need of determining how to worry most effectively about the really important problems."

Maddox would also like to see scientists and their organizations become involved in providing information on topical and sensitive socio-scientific issues directed both at school children and at politicians and other opinion leaders. "The aim of approaches towards the latter should not be aggressive lobbying, but to provide well-prepared materials and to hold meetings presenting clear, simple information. Pressure groups indulge in this sort of propaganda all the time. Why don't we use the same type of opportunity?"

An example of the type of sensitive topic which Maddox feels strongly about, is the use of animals for research and toxicology screening. He believes that this practice must continue in the interests of public health and is highly critical of those sectors of the scientific community that have not spoken out on the subject. "Many people have been mealy-mouthed on the issue of animal rights, and have not publicly said what they really think. They have consequently done much to damage their own cause."

Regarding the use of information in research, Maddox feels that electronic databases have not yet made their full impact. He is certain, however, that they will soon make it far more convenient for researchers to secure the information they require on a daily basis. "One particularly important aspect to consider is what can be done to provide information that will simplify the use of existing technical devices. Take computers, for example. There is no reason why we should have to spend two days mastering a new piece of software when it is bought. Why can't we dial up a computer that quickly teaches us how to use the new software? Much more should be done to provide electronic skills by means of automated teaching aids. Otherwise, there's no way of properly exploiting the software which is now being developed and introduced."

'Nature' already uses electronic communication to deal with authors and Maddox predicts that this will increase considerably in future. "But we will all continue to read paper journals simply because they are more convenient than using a computer screen. Also don't forget that while computer technology is becoming speedier and more versatile, there are parallel advances in conventional printing technology. We now print 'Nature' in four centres, including Japan and China, and I wouldn't be surprised if we are printing in the Soviet Union and the Third World too by early in the next decade."

American scientific enterprise

Maddox believes that science is in particular good health at the moment in the United States, especially in pure and applied molecular biology and research into molecular and atomic structure and its applications in such fields as spectroscopy and lasers. "Why this is so is a strange business," explains Maddox. "American investment in basic science grew during the Reagan Administration at a time when it was stagnant elsewhere, or was in fact declining in effective terms, as in Britain. To some degree this was because the US Administration was not worried about spending money at that time, whether it was on pure science, SDI or social security programmes."

"Such support also indicates that fundamental research was seen as an important basis for industrial competitiveness. My own belief is that although basic science is a necessary condition for economic growth, it is insufficient by itself. Nevertheless, the Americans have learned how to apply entrepreneurship in conducting research to such a degree that the success of a university depends on the ability of its scientists to win grants. The paradox is that this situation can not be sustained by American citizens alone, so that there's now a brain drain from Europe and India, as well as from the Soviet Union, which is very concerned about the loss of scientific skills. Another paradox is that the competitiveness of the American system, particularly the belief that success is determined by how much and where research is published, doesn't encourage cooperation between scientists."

Maddox also believes that American industry has gone 'off the boil' recently because of a shortage of able technical managers at the executive level. "They lack imaginative team leaders in high technology, so that when they do embark on daring projects nowadays, such as the Space Shuttle, they tend to be unwilling to take risks. Everything is controlled by stultifying committees, so that decisions are made too slowly, and dangerous compromises are reached. The result is that the United States is now moving less quickly than Japan, although the difference is only marginal."

Science in Western Europe

"The overall healthy situation of American science is good for science elsewhere too," Maddox continues. "In terms of discovery, Europe is also faring better than it was ten years ago, with the exception of Britain. And the Japanese are rapidly learning about the importance of basic research. This is reflected in the fact that the trickle of papers we publish from Japan is now much on the increase."

Maddox goes on to describe how the lack of competent industrial middle managers in Britain relates to the lowly status of engineers in that country. "From the minute they join a company, engineers are undervalued," he says. "They are badly paid and soon discover that the only way they can advance their careers is to move into other spheres. As engineers, they are simply not given the opportunity to contribute as much as they might to their company's prosperity. By contrast, France has been staggeringly successful in the way in which people of similar standing have been used to perform projects which have transformed the country. The French telephone system, for instance, was seen as a joke as recently as twenty years ago. But just look at it today. You can now call anywhere in the world from a coin box on the Champs Elysées."

"In the case of the Federal Republic of Germany, science has been generously backed over the past two decades. I think that the fruits of this investment are now beginning to show in the academic field, especially in physics and mathematics and increasingly in biochemistry and molecular biology. Again, their industrial success stems from the intelligence with which they use their technical management. But I'm concerned about the growth of the anti-science movement in West Germany which seems to be related to a legal system that provides many opportunities for dissenting groups to block industrial and other developments."

"As for The Netherlands, the way in which the country was crammed with outstandingly important technical enterprises such as Philips, Shell and elsewhere some twenty years ago was very impressive. This activity seemed certain to grow stronger, but it hasn't. I suppose the reason is that unlike the Americans, the Dutch have been unable to recognize that they need to significantly depend on technically-qualified people from elsewhere."

Science and ideology

Commenting on events in Eastern Europe, Maddox believes that it is important to distinguish the Soviet Union from other countries.

"The scientific enterprise has been virtually destroyed in some countries such as Romania, as a result of free thought being suppressed by the government. By contrast, the shape of things is not too bad in East Germany, Czechoslovakia and Poland. The best strategy now would be for the six newly liberated nations to form a collaboration and cooperate, instead of having old-style Stalinist academies in each country."

"I believe that, in time, they will force themselves on the West's attention, which will itself wish to collaborate with their most creative researchers. I envisage a lot of joint research arising in this way. As a contribution to cross-fertilization, 'Nature' has in fact recently begun a scheme allowing Eastern European scientists to receive the journal free of charge for one year."

"The Soviet Union, despite its Stalinist past, has exceedingly able scientists who are still practising good science, and the country could become a powerful force very soon. A major problem at the moment is that alongside the excellence, Soviet institutes have too many people doing poor-quality work. I think it would be very beneficial for Soviet science if the budgets of all institutes were reduced to a third of the present figures. The incentives to improve the overall quality of science in the Soviet Union are certainly enormous. What they need to do now is to modernize agriculture and industry, and manufacture computing equipment which can not be bought in the West. I predict that they will succeed and that this will in turn benefit science throughout the world."

Maddox is encouraged by the recent events in Eastern Europe, but is concerned about the growth of other ideologies. "We have just been through a period when the Marxist world has confessed its impotence, and other ideologies such as Islam have not had a very good record when it comes to science. What both science and society surely need in future is ideology-free societies. That's why I welcome the fact that in countries where ideology doesn't run too deeply, it has been possible to devise legislation that satisfies all sides on such issues as in-vitro fertilization. But some of the experiments in democracy we're now seeing in Eastern Europe differ from our Western models by allowing a greater degree of public participation at all levels. I feel that they may provide interesting models for the way in which we should also run our countries."

Bernard Dixon

ABOUT THE PEOPLE INTERVIEWED...

David Halberstam

Writer and journalist David Halberstam was born in New York in 1935. After graduating from Harvard College in 1955, he worked as a reporter on several newspapers in the USA including the Nashville 'Tennessean' for a period of four years. As foreign correspondent for the 'New York Times', he spent six years abroad in the Congo, Vietnam and Poland. He had written several books before he gained a reputation as a critical observer of modern times with his book 'The Best and the Brightest'. This study of the US Government and military in the Vietnam era was followed by other major works. 'The Powers that Be' is concerned with the rise of the modern media and 'The Reckoning' is a comparative history of the Ford and Nissan automobile companies. His most recent work is 'The Next Century', about the year 2000 and beyond.

Seun Ogunseitan

Born in Nigeria in 1960, Oluwaseun Oladapo Ogunseitan studied zoology at the University of Ibadan. He joined the staff of the Nigerian newspaper 'The Guardian' as science correspondent in 1984. Specializing in reporting on environmental issues, Ogunseitan won the Nigerian Journalist of the Year award in 1987 for his story about the pollution caused by a new chemical complex at Port Harcourt. During a visit to the USA in 1988, his ideas crystallized into a specialized information agency. Returning to Nigeria, he resigned from 'The Guardian' to set up the African Centre for Science and Development Information (ACSDI).

Federico Mayor Zaragoza

Federico Mayor Zaragoza was born in Barcelona, Spain, in 1934. After studying pharmacy in Madrid, he became a researcher in biochemistry and was appointed Professor of Biochemistry at the University of Granada in 1963. As the new Rector of the University in 1967, he was automatically granted a seat in Parliament. In 1978 he was appointed Deputy Director General of Unesco, and returned to Spain in 1981 to become Minister of Education and Science in the democratically elected Government of Adolfo Suarez and then a member of the European Parliament. Since 1987 he has been Director General of Unesco. He is the author of several books, the latest being 'The New Page'.

Alexander King

Founding member of the Club of Rome, Alexander King has been its President since 1984. Born in Glasgow, Scotland, in 1909, he was educated at the Imperial College in London, where he obtained his doctorate in chemistry; subsequently he was lecturing there until the outbreak of the Second World War. As Deputy Advisor to the Minister of Production, he was closely involved in the development of an effective insecticide, DDT. Continuing his career in the British civil service after World War II, he became Chief Scientist of the Department of Scientific and Industrial Research. Next, he went to Paris as Co-Director of the European Productivity Agency. In 1961 he was appointed Director for Scientific Affairs of the newly founded Organization for Economic Cooperation and Development and later became its Director General for Science, Technology and Education.

Erich Bloch

Erich Bloch, born in Germany in 1925, has been Director of the National Science Foundation in the United States since 1984. Before joining the NSF, he served as Corporate Vice President for Technical Personnel Development at IBM, where he had worked since 1952, originally as an electrical engineer. From 1981 to 1984 he acted as Chairman of the Semi-Conductor Research Cooperative, a group of leading computer and electronics firms that support advanced research in universities. Bloch was awarded the National Medal of Technology by President Reagan in 1985 for his contribution to pioneering developments in the computer industry.

Harry Beckers

Top scientist and Group Research Coordinator for Royal Dutch/Shell, Dr. Harry Beckers was born in Maastricht, The Netherlands, in 1931. After gaining his doctorate in technical sciences at the Delft University of Technology, he started his distinguished career at Shell as a research physicist. His first promotion was to Head of the Physics Division at the Royal Dutch/Shell Laboratory in Amsterdam, and subsequently he became Head of the Physics and Mathematics Division. In 1970 he moved to London, England, to act as Head of the Shell Planning and Strategic Studies Division and later of the Organization Services Division. He returned to The Netherlands in 1977 to become Group Research Coordinator of the oil company.

Etienne Davignon

Viscount Etienne Davignon was born in Budapest in 1932. The son of a Belgian diplomat, Davignon read law and took his doctors degree at Louvain before following family tradition and becoming a diplomat. He served abroad for several years before returning to Brussels to become Head of the Cabinet of Foreign Secretary Paul-Henri Spaak, and later General Director of Political Affairs. He was appointed as the first Chairman of the International Energy Agency in Paris in 1974, and three years later began the first of two four-year terms as a Commissioner of the European Commission. He left the Commission in 1985 to take over responsibilities as Manager of the Société Générale de Belgique, a large commercial and industrial holding company that dominates the Belgian economy, and was named its Chairman in 1988.

Robert Solow

Robert Solow, Nobel Laureate and one of America's foremost economic theorists, was born in Brooklyn, New York, in 1924. He graduated from Harvard University in 1949 and has served as advisor to two US presidents. Currently a researcher at the Massachusetts Institute of Technology in Boston, he was appointed Vice-Chairman of MIT's Commission on Industrial Productivity, which released a provocative report in May 1989 on the problems distressing the American economy. He was elected to membership of the US National Academy of Science in 1972 and awarded the Nobel Prize for economics in 1987.

Hisao Yamada

Professor Hisao Yamada, born in Japan in 1930, studied electrical engineering at the University of Tokyo until 1953, whereafter he continued his studies in the USA at the University of Pennsylvania; there he obtained a doctorate in computer and information science. In the first half of the sixties he worked first as a researcher with General Dynamics and later with the IBM Corporation, both in New York. In 1966 he returned to the University of Pennsylvania as an Associate Professor of Computer and Information Science. His position there as Vice-Chairman of Graduate Studies ended with his return to Japan. Since 1972 he has been Professor at the Department of Information Science of the University of Tokyo. In between, he spent an additional two years in the USA, at the University of Delaware and Stanford University. At present, he is the Director for R&D at the National Center for Science Information System (NACSIS) in Japan. Professor Yamada is the author of several research papers.

Tudor Oltean

Tudor Oltean is Professor of Communication Sciences at the University of Amsterdam, The Netherlands. He was born in Timisoara, Romania, in 1943 and educated in Bucharest, Romania. Between 1966 and 1777 he was Assistant Professor at the Department of Comparative Literature, University of Bucharest.

In 1973 he published his thesis, entitled 'The Morphology of the 18th Century European Romantic Novel', which slipped past the censors despite its non-Marxist interpretation. All of his work was later banned in Romania after his refusal to return to Bucharest from an exchange visit of scientists to Amsterdam in 1977. He obtained a post at the University of Amsterdam, successively at the Department of Roman Languages, the Department of General Literature and the Department of Communication Sciences. His major works deal with the morphology of the European novel in the 18th century (published in 1974), and in the 19th century (1978) and with the mediatization of narratives in contemporary culture (in preparation).

Rudolf Bernhardt

Professor Rudolf Bernhardt, born in Germany in 1925, studied at Frankfurt University and obtained a doctorate in jurisprudence in 1955, specializing in comparative public and international law. In 1954, he began work as a research assistant at the Max Planck Institute of Comparative Public Law and International Law, later becoming a research fellow. Bernhardt was named as Director of the Institute in 1970. Since 1981, he combines this position with duties representing the Federal Republic of Germany in the European Court of Human Rights in Strasbourg.

Roger Penrose

The scientific achievements of Roger Penrose, Professor of Mathematics at The University of Oxford, England, include his contribution to a theory explaining the presence of Black Holes in the Universe, which was formulated in 1965 together with Cambridge Professor Stephen Hawking. In 1974, he discovered pairs of mathematical shapes known as the 'Penrose Tiles', which proved to have unexpected significance for materials science. He recently published a book for general readership about Artificial Intelligence, 'The Emperor's New Mind', which deals with a wide range of topics to present a strong argument against the widespread belief that computers in their present configuration could take over human thinking.

Kai Siegbahn

Born in Lund, Sweden, in 1918, Kai Siegbahn is Emeritus-Professor of Physics at the University of Uppsala. After studying physics, mathematics and chemistry, he chose the new field of nuclear physics for his doctorate. Discoveries from extensive investigations lead to the publication in 1967 of the book entitled 'Electron Spectroscopy for Chemical Analysis'. In 1981, he was awarded the Nobel Prize for physics for his development of electron spectroscopy. Since his retirement, he has continued to work at his laboratory at the University of Uppsala on the combination of lasers with electron spectroscopy. Siegbahn is involved in The World Laboratory, an organization providing funds and scholarships for projects in various countries.

Shigeo Minowa

Shigeo Minowa was born in Tokyo, Japan, in 1926. After graduating from Tokyo University in 1950, he worked as Managing Director of Tokyo University Press. In 1972 he was elected President of the International Association of Scholary Publishers (IASP), for a period of four years. After having worked for the Tokyo University Press for 24 years, he moved to the United Nation's University in Tokyo, as Head of their Scientific Information Resources. From 1985 until the beginning of 1990 he was a professor at the International Communication Department of the Aichi Gakuin University. Since then Professor Minowa has been with the Kanagawa University, Tokyo, as Director of the Institute of International Business and Management.

John Maddox

John Maddox has been Editor of the science journal 'Nature' since 1980, a position he held previously between 1966 and 1973. In 1955, he had left a career in theoretical physics at the University of Manchester to become science correspondent of the newspaper 'Manchester Guardian' for some nine years. Maddox was appointed as Managing Director of Macmillan Journals Ltd. in 1970, before acting as Chairman of Maddox Editorial Ltd. in 1972 for two years. During his career, he has been an affiliate of the Rockefeller Institute in New York (1962–1963), Director of the Nuffield Foundation, London (1975–1980), and a Public Interest Representative on the UK Genetic Manipulation Advisory Group (1976–1980). His books include 'The Spread of Nuclear Weapons' written with Leonard Beaton (1962), 'The Doomsday Syndrome' (1972) and 'Beyond the Energy Crisis' (1975).

Index